Praise for *Advanced Top Bar Beekee*

Anyone who encourages better handling of nature's number one livestock—bees—should be read and followed. Christy Hemenway is that guy. Read and learn.

— Joel Salatin, Polyface Farm

Christy's experience and drive to further the use of top bar hives is extremely evident in this her next level work. Her first book got you into a hive, why it is a good thing to do, and how to make it work. I learned a few tricks from her and my top bar beekeeping improved due to her insights and explanations. But what about next year? That's where this work picks up. It gets you through winter, spring, swarms, feeding, splits, harvesting honey and then settles into the very best thing I can say about this form of keeping bees. Clean wax. If for no other reason do you begin or continue keeping bees than the fact that you have clean wax with these hives… then you have made the very best decision you can.

—Kim Flottum, editor, *Bee Culture* magazine,
and editor, *BEEKeeping: Your First Three Years*

I was amazed to find that what happened in my first top bar hives was exactly what Christy had described in her first book. Her new book is not only essential for those who want to keep bees in top bar hives, but also for those want a deeper look on beekeeping problems and on the life of Apis mellifera.

— Paolo Fontana
Entomologist / Apidologist

The loss of biodiversity on our planet is one of the greatest emergencies of our time. We can't wait any longer: several thousand species become extinct each year! Human beings must be involved in conservation of ecosystems and biodiversity. In this struggle against time, all the beekeepers can play a fundamental role. They can become the vanguard of a movement of responsible citizens involved in biodiversity conservation all over the world. And, at the head of this vanguard, persons like Christy Hemenway can have a leadership role in building a better future for all!

—Gianfranco Caoduro, Honorary President, World Biodiversity Association

Here are your next steps to keeping bees in top bar hives. Thoughtful, experienced, articulate advice.

— Michael Bush, BushFarms.com

Good ideas are just good ideas unless you can get a significant number of people over a sizable geography to buy into them. Christy has done just that! *Advanced Top Bar Beekeeping* is a treasure. For the top bar beekeeping disciples that Christy has already launched this book will support sustainability and serve to draw more practitioners into the adventure. Congratulations Christy! You have made a significant contribution to the top bar craft!

— Marty Hardison, Top Bar Pioneer

advanced TOP BAR Beekeeping

advanced TOP BAR beekeeping

NEXT STEPS for the *thinking* beekeeper

CHRISTY HEMENWAY

Printed in Canada. First printing January 2017

Inquiries regarding requests to reprint all or part of *Advanced Top Bar Beekeeping* should be addressed to New Society Publishers at the address below. To order directly from the publishers, please call toll-free (North America) 1-800-567-6772, or order online at www.newsociety.com

Any other inquiries can be directed by mail to:

New Society Publishers
P.O. Box 189, Gabriola Island, BC V0R 1X0, Canada
(250) 247-9737

LIBRARY AND ARCHIVES CANADA CATALOGUING IN PUBLICATION

Hemenway, Christy, author

Advanced top bar beekeeping : next steps for the thinking beekeeper / Christy Hemenway.

Includes index.
Issued in print and electronic formats.
ISBN 978-0-86571-809-8 (softcover). — ISBN 978-1-55092-605-7 (ebook)

1. Bee culture. 2. Honeybee. 3. Beehives. I. Title. II. Title: Top bar beekeeping.

SF523.H44 2017   638'.1   C2016-907508-7
C2016-907509-5

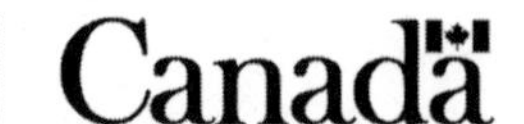

New Society Publishers' mission is to publish books that contribute in fundamental ways to building an ecologically sustainable and just society, and to do so with the least possible impact upon the environment, in a manner that models that vision.

Contents

Acknowledgments

First, thanks to Geoff Keller. I know he didn't know what he was getting himself into when he found himself living with this author as she completed her second book – but his patience, support, and encouragement were crucial in seeing this project progress from start to fruition. Thanks are also due to Helen Fenske of Kentucky who was instrumental in helping move things forward.

I was deeply honored when *The Thinking Beekeeper* was translated into Italian, and upon its publication, I was invited to travel to Italy to present the book and to speak to beekeepers at several meetings. Important work is being done in Italy concerning beekeeping in top bar hives in support of biodiversity, led by Paolo Fontana, an entomologist at the Edmund Mach Foundation, and the president of the World Biodiversity Association. Paolo describes their work:

In Europe *Apis mellifera* is a native species, and it is fundamental for the conservation of biodiversity. In 2015, the World Biodiversity Association (biodiversityassociation.org), and the Edmund Mach Foundation (fmach.it) launched the project Bees for Biodiversity to recreate the network of wild colonies of *Apis mellifera,* that was lost following the arrival of varroa destructor.

Bees for Biodiversity aims to encourage the spread of backyard beekeeping with top bar hives, a type of beekeeping hitherto almost unknown in Italy. The first action was to bring together a group of volunteer beekeepers for a shared experience to define a top bar model and to examine a great deal of literature. The new model was named BF top bar, where BF means Biodiversity Friend, a brand owned by WBA.

A second fundamental step was the publication by WBA of the Italian edition of *The Thinking Beekeeper*, which greatly increased the interest in top bar beekeeping, being the first book in Italian on this topic.

A year and a half later, more than 500 top bar hives have been built and populated, thanks to this project. At the Edmund Mach Foundation, there are now more than 30 top bar hive colonies being used to study the management of bees in these hives and for bio-monitoring in cultivated areas, to study contamination and fitness of honey bees. Much has been accomplished in a short time, and the interest continues to grow.

So sincere thanks to Paolo Fontana and Gianfranco Caoduro, president and honorary president, respectively, of the World Biodiversity Association; the Edmund Mach Foundation; Sara Marcolini, who did a magnificent job with the translation into Italian and the many kind and generous Italian beekeepers I met on my trip to Italy in 2015. Their work has informed my own, widened my own horizons, and buoyed my hopes for a world that supports the preservation of biodiversity, a world with more organic food, less industrial agriculture, and more natural, less invasive, top bar hive beekeeping.

Introduction

The Thinking Beekeeper Thinks Again

There was some discussion about calling this book *The Thinking Beekeeper Thinks Again*. I thought that might appeal to those of you who had read the first book and found it worthwhile—partly because this would be an extension of the original mindset, and it would offer more of my beekeeping philosophy and management techniques. Besides, it was kind of funny, and I thought most of you would get it.

But the world has evolved to where we now live with some terrible, tragic things:

- a profound disconnect from nature and natural systems
- a worried realization that everything is connected; when we harm a piece of the planet, or damage one of its systems, we truly harm all of it, and all of its inhabitants
- the knowledge that we have, indeed, already done this harm, perhaps past the point of no return

This can be really hard to live with on a day-to-day basis. It's frightening. It's depressing. It leaves parents worried about the world their children will live in—or be unable to live in—about whether there will be a world left for anyone to live on seven generations from now. Sometimes it can be just too hard to keep trying, to keep believing that you can make any difference at all, and we sometimes come to a point where we do "think again"—but in that case, it might mean that we give up on something important, we quit trying, and we regret all the effort we have invested thus far.

But I didn't want anyone to give up, and I certainly don't want anyone to regret their decision to keep bees, even though it's fairly complex nowadays, for all the reasons discussed in this book and more. So instead, the official title became *Advanced Top Bar Beekeeping: Next Steps for the Thinking Beekeeper*.

Around Gold Star Honeybees' Global Headquarters, we still call it *The Thinking Beekeeper Thinks Again*, or *TBx2*. We sort of suspect that you might too.

Questionnaire

How much poison are you willing
to eat for the success of the free
market and global trade? Please
name your preferred poisons.

For the sake of goodness, how much
evil are you willing to do?
Fill in the following blanks
with the names of your favorite
evils and acts of hatred.

What sacrifices are you prepared
to make for culture and civilization?
Please list the monuments, shrines,
and works of art you would
most willingly destroy.

In the name of patriotism and
the flag, how much of our beloved
land are you willing to desecrate?
List in the following spaces
the mountains, rivers, towns, farms
you could most readily do without.

State briefly the ideas, ideals, or hopes,
the energy sources, the kinds of security;
for which you would kill a child.
Name, please, the children whom
you would be willing to kill.

—Wendell Berry,
from *Leavings: Poems,* Counterpoint, 2010

PART I

Year Two: What to Do?

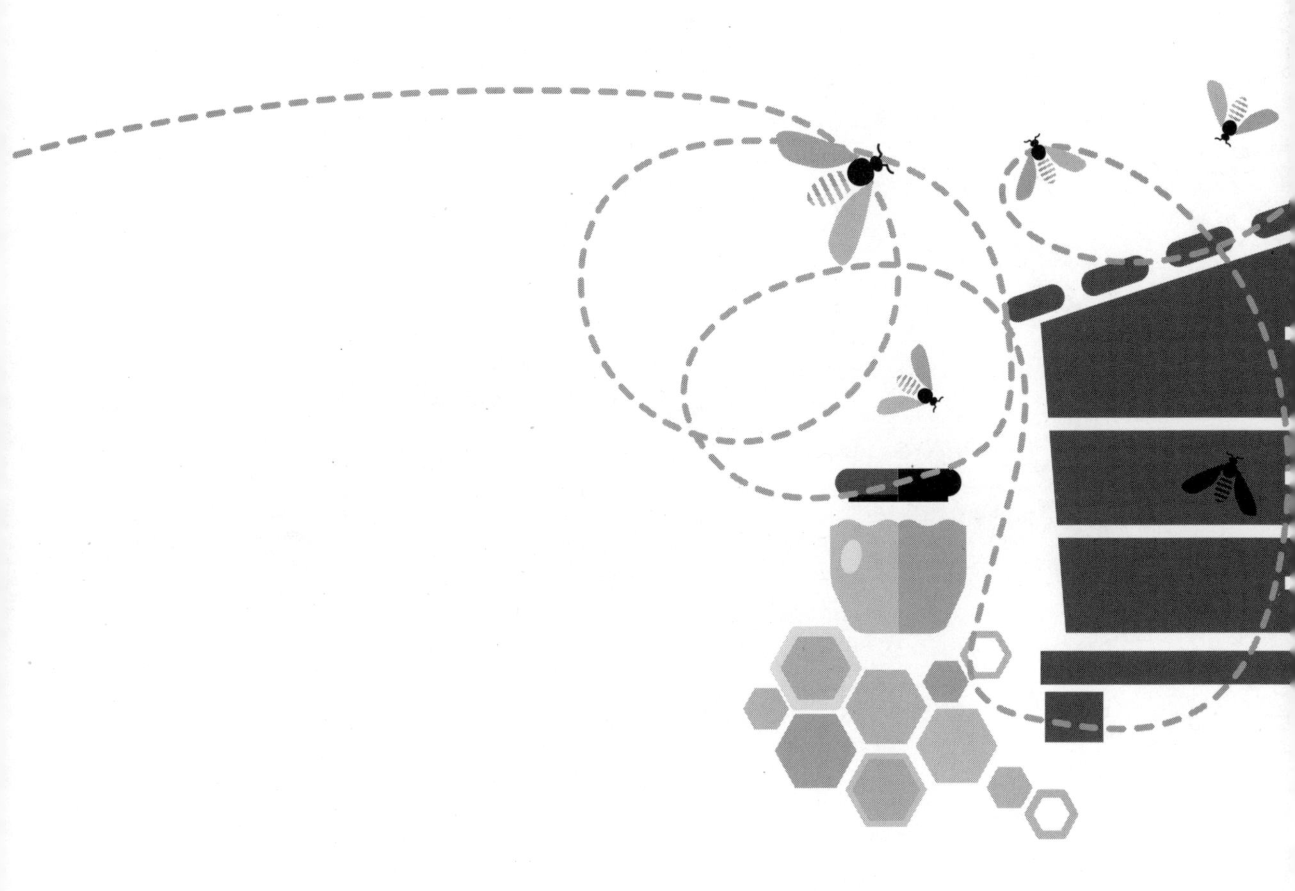

1 Year Two: How Did We Get Here?

At the risk of repeating myself, I thought about titling this chapter "How Did We Get Here from There?" just as I had for Chapter 1 of *The Thinking Beekeeper*. But in that chapter, we fast-forwarded from the honey hunters of ancient civilization to managed beekeeping with fixed comb hives in Egypt and the unspoiled honey found in King Tut's tomb; then we moved on to the "Greek Beehive," generally described as the original top bar hive; and from thence to the Reverend Lorenzo Lorraine Langstroth and his movable comb, square-box beehive—a revelation to antebellum America. All this to understand why we do what we do in beekeeping today—using movable comb in managed hives.

It was a relatively short hop from the Reverend Langstroth's patent in 1853 to the post-World War II era, and the drive to scale up and mechanize our food system, which industrialized beekeeping right along with it. Former Secretary of Agriculture Earl "Rusty" Butz and his "Get big or get out!" philosophy threw us headlong into the perpetual imbalance of large-scale monoculture agriculture, with its synthetic fertilizers and unrelenting applications of toxic pesticides. The law of unintended consequences soon revealed our shortsightedness about the sustainability of a system forced so far out of balance.

FIGURE 1.1. **Zoom in to the bee yard.** Credit: Harry Kavouksorian.

Today we want to narrow our focus, leaving behind ancient history, the Industrial Revolution and the cares of the modern world, zooming in really close—focusing on *your* bee yard in its second spring.

Just how *did* we get here?

At the end of Year 1, you probably invested some time and effort in preparing your top bar hive(s) to withstand winter's ravages—from careful placement of the hive before you even installed your bees, to providing protection via insulation, hay bales, tarpaper, rigid foam or another resourceful solution that minimized the effect of the winter wind. You likely managed them in such a way that the bees built their combs in one direction, and their food stores were located on only one side of the *brood nest* (glossary includes terms in italics), not on both sides. It's likely that you were extremely conservative about harvesting honey, if indeed you harvested any at all. Or maybe you did remove a bar or two, and left the honey on the bar as a reserve for feeding back to them later if needed. (Note that this will require having a few spare top bars so that your hive can always have the full complement of bars, even if you have removed some of them.) Then in the fall, you put your bees to bed, knowing that they were on their own inside the hive.

For beekeepers, winter can be hard. Not just because the weather is seriously cold and snowy, or at least drab and dismal, but because you can't check on your bees! It's a good time for inside activities, such as building new hives, rendering wax, devising new gadgets, sorting through last year's pictures, catching up on your reading—but all the while there's a bit of an itch—wanting to know what's going on in the hive. Do they have enough honey? Are there enough bees? Is the hive strong enough to make it through what passes for winter where you are? What is happening in there!?

Then, around mid-January, we get one of those days that completely restores our belief that spring will indeed come. The temperatures rise into the mid-50s Fahrenheit, and lo and behold—there *are* still bees in that box!!! And out they come for the storied *cleansing flight*—leaving orangey-brown spots on the snow, on your car windshield and on the sheets you hung out to dry because it was just too beautiful to run the dryer. (Hey, it's a long time to go without going to the bathroom...!)

FIGURE 1.2. Top bar hives in winter. Credit: Christy Hemenway.

What a heartening sight, that bee poop! Now you know they're alive... but it's terribly early still. There may be months to go before the weather truly breaks, and the local early nectar sources begin to *bloom*. You may be worried about their food stores. If supplementing their honey stores mid-winter is part of your paradigm, this is the kind of day to check on their stores.

Because opening the hive when the bees need to be clustered is not something to do casually, be sure to get your ducks in a row first—before you crack open the box, breaking their propolis seal and disrupting their winter cluster.

What do you feed honeybees in winter? The syrup feeder you used last spring is not an option now—for one very important reason: It's too cold! To survive cold temperatures, bees must *cluster*. An individual bee at temperatures below 45°F becomes paralyzed and cannot even return to the cluster. The syrup feeder will be located much too far away from the cluster for the bees to access it. So perhaps it's more a question of *where* can you feed honeybees in winter? Bees clustered for winter need to be touching their food source—like they would if they were clustered on full, ripe *honeycomb* made of natural wax. You may have some bars of honey in reserve—and that's the ideal food for bees.

How to Make Fondant for Winter Bee Food in Your Top Bar Hive

Notes

1. You will need a candy thermometer; this recipe is temperature sensitive!
2. Do not use raw, turbinado, beet, or brown sugar. Organic cane sugar is fine. Read the label closely. If it doesn't say cane sugar, it is probably beet sugar.
3. 2¼ cups of sugar weighs 1 pound.
4. When made using 1 cup of water to 4 cups of sugar, this recipe will fill a typical fondant feeder frame.

How to Make the Fondant

Combine:

- 1 part water to 4 parts sugar
- ¼ teaspoon of vinegar per pound of sugar (this helps break down the sugar)
- ¼ teaspoon of salt (preferably a salt containing beneficial minerals)

1. Bring to a boil, stirring constantly until the mixture boils. (Very important!) Cover and boil for 3 minutes WITHOUT stirring. Continue to boil until the temperature reaches 234°F. (Exceeding this temperature will caramelize the fondant, which is harmful to bees.)
2. At 234°F, remove the mixture immediately from the heat and allow it to cool to 200°F.
3. Meanwhile, arrange your fondant feeder frame on a flat surface covered with waxed paper. Put the thicker edge of the top bar over the edge of the flat surface, so that the frame itself lays flat and works to contain the fondant.
4. At 200°F, use a whisk to whip the mixture until it turns white.
5. Quickly pour the mixture into your feeder frame. Allow the fondant to cool completely. Remove the waxed paper.
6. The fondant feeder can then be stored in the freezer in a plastic bag.
7. If you determine that you need to supplement your bees' natural honey stores, place the fondant frame in the hive beyond any existing bars of honey so that they first devour their own honey stores before moving into the fondant frames.

FIGURE 1.3. Commercial fondant is available in larger quantities. Check the ingredients list!
Credit: BeeCurious on BeeSource.com.

But if we are assuming you don't have any of your bees' own honey to feed back to them, then the food of choice would be *fondant*. As a solid, fondant adds little moisture to the hive, and it does not require a distant feeder jar in order to work. Fondant can be hung from the top bar, imitating a comb filled with honey, and this top bar can be placed right next to the cluster of bees. This is crucial, since bees must stay in cluster to survive. They cannot survive away from the cluster, nor can they move as a cluster across empty comb to get to a distant food source. This is why it sometimes happens that there may be plenty of honey in the hive, but the bees can starve if it is not located where they can get to it.

So you're prepared for the eventuality that your bees may need food when you check them. You've got your bee gear on, and now you're ready to go look. You're ready with a bar of honey or fondant if they do need food. If they don't, you can just grin and close up.

Here's how to check. Begin at the honey end of the hive. There is no reason to start in the brood nest; it is too cold to tear into it, and you are only here to check food stores. Once you have opened the hive, starting at the honey end and moving toward the brood nest, click through bars until you come to bars of either bees or honey. If you run into honey without seeing any bees—great! You don't need to feed them! But, if you run into bees first—place your bar(s) of fondant or honey right there next to them. Put the rest of the bars back in, close up and cross your fingers. It's still a long way to spring!

So let's fast-forward another little bit. You probably did that food check in January, possibly February. If you live in the South, your bees will likely be all set with natural food sources shortly, and they'll be humming along. But perhaps where you live, the growing season doesn't really get started until closer to April, or even May. So your bees have got more hanging on to do.

Northern or cold winter bees are doing an amazing balancing act inside their hive at this time of year. Believe it or not, the queen begins laying eggs while it is still quite cold, and the bees need to cluster over those new baby bees, to warm and feed and protect them, especially through the egg and larvae stage. Meanwhile, the older bees—the ones that were

FIGURE 1.4. A fondant feeder must suspend the fondant where the bees can access it.
Credit: Christy Hemenway.

FIGURE 1.5. This fondant feeder supports the fondant with ½-inch hardware cloth.
Credit: Maury Hepner.

born in fall and are anatomically different so they can live through a cold winter but did very little foraging—are dying off pretty rapidly. New bees are being born. Nectar and pollen may or may not be scarce, thanks to whatever the weather is doing.

Welcome to April, known for being "the cruelest month in beekeeping." How heartbreaking it can be to see bees in January, and in March perhaps, but then, thinking all is well, discover in April that the hive is dead. It's entirely possible that a hive may *overwinter*, but then not *overspring*. (And yes, sometimes we make up words in beekeeping to get across what we mean...)

It's likely that the new beekeeper is doing one of two things at this point: Either celebrating success or mourning a dead-out hive. Of course we would much rather see a thriving hive come through a cold winter with flying colors, but if this was your first winter, and you are mourning the loss of your first hive, fear not—all is not lost!

Consider where you were last spring, such a short time ago, with an empty top bar hive, waiting for your first bees. Today you have some precious things: a year's worth of knowledge and experience; and something else—bars of naturally drawn beeswax honeycomb, wax that was made by bees, for bees. If you did not use any toxic treatments in your hive, then your natural wax comb is incredibly clean, containing only those environmental toxins that were brought in during the season by your foraging bees.

One of the most important things you can do in the way of letting bees act in accord with their natural systems is to leave the making of the wax entirely up to them. Commercially available wax *foundation* has been tested and found to be contaminated with the persistent pesticides and chemicals that were purposely introduced into the hive by beekeepers, and plastic foundation is,

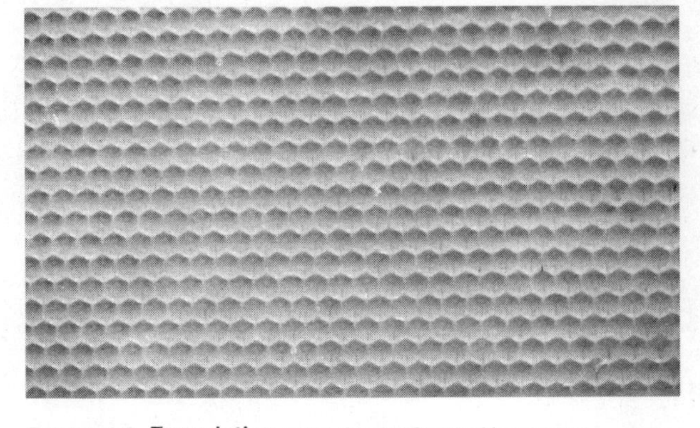

FIGURE 1.6. Foundation wax. Credit: Christy Hemenway.

well, plastic. You've probably heard me say this before, but the truth bears repeating: For the bees, "It's all about the wax."

Because wax building is resource-intensive for a new package of bees—and, of course, there are a million other variables and opportunities for failure in the first year of a top bar hive—it's really worth appreciating just what your bees actually did accomplish, even if they did not overwinter. They left behind this precious gift of natural comb. So yes, let's mourn the loss of your bees, but let's also celebrate this new and very important resource.

FIGURE 1.7. Fresh white comb. Credit: Dana Gray.

Let's take a look at these two different scenarios and plot a path forward for this new season.

If your hive succumbed to the stresses of winter, your next steps are fairly straightforward, even if somewhat uninspiring. They include cleaning dead bees out of the hive, diagnosing any disease symptoms that might exist and resetting for new bees.

FIGURE 1.8. Freshly capped honeycomb. Credit: Christy Hemenway.

Note that this task list *does not include* destroying the existing comb! You may be uncertain about the condition or cleanliness or safety of reusing the comb, since now it looks so much different than when it was first drawn by the bees. Fresh, white, beautiful *honeycomb* is a long way from dark brown *brood comb* that may even contain bee carcasses. So what should it look like?

FIGURE 1.9. Brown brood comb.
Credit: Christy Hemenway.

Beeswax comb starts out white, and it is very soft and extremely fragile. It has just been excreted fresh from the wax glands in your bees' bellies. Over time, things happen that alter this ethereal beauty: the wax hardens, it oxidizes, it yellows, it turns brown. The bees walk on it, staining it with pollen, nectar and propolis resins. They store honey and pollen in the cells, and the color changes again, affected by the elements that make up the honey and pollen. But brood comb changes the most; it actually turns brown. This is due to the part brood comb plays in the life cycle of the bees—that magical moment when the bee pupae, capped inside their tiny hexagonal cells, spin a very thin silk cocoon around themselves as they begin their metamorphosis from grub-like larvae to newly hatched honeybees. When the bees hatch from their cells, these cocoons are left behind, causing the brood comb to become dark brown. We humans tend to associate brown with dirty, but brown does not equal dirty in this case! Brown is good, and from the bees' point of view, the browner the better. Brown brood comb is the perfect *anchor* for a new hive of bees. It also works beautifully for luring a swarm into your *swarm trap* should you decide to do that. In any case—don't destroy the natural wax comb your bees have made.

If you have doubts about disease issues, consult your state apiarist, local extension agent and/or and experienced beekeepers. You don't want to perpetuate a disease problem. But you sure don't want to waste all the hard work done by your first year's bees either.

Another opportunity is also at hand: It's much easier to make a change to your hive when there are no bees in it! Take advantage of this time to resolve any design issues you found with your hive. Changes to bottom boards, entrances, roofs, legs and observation windows are all much simpler to implement with no bees in the box! If you should decide to make a major changeover in your top bars, to which the comb is attached, again, *do not destroy* your first-year bees' comb. It can be removed from the old top bars and rehung from new ones. The bees are quite capable of reattaching it to the bar.

Here's how to rehang the combs. (You may recognize this from page 87 in *The Thinking Beekeeper: A Guide to Natural Beekeeping in Top Bar Hives.*) Carefully cut the comb from each of your old bars. Try to cut a reasonably straight line across the top edge, as you will want to have as straight a surface as possible touching the new bars' comb guides.

Then for each bar that needs to be rehung:

- Cut two ½-inch wide strips of some slightly stretchy fabric. (T-shirt material works well for this.) They need to be a little more than twice as long as your comb is tall, or twice as tall as your hive is deep.
- Have on hand 4 flat-head thumbtacks. They work better for this than push pins because they do not stick out so far as to get in the way of the top bars when you put the bars back into the hive.
- Tack two fabric strips to the underside of the top bar. Turn the bar over on your work surface, placing it with the tacks down and the fabric strips extended toward you.
- Lay the piece of comb down on top of the fabric strips, as near to the top bar as possible.

- Bring the other ends of the fabric strips up and over the comb, and attach them to the underside of the top bar as well, using the other 2 tacks. Adjust if needed so that the comb is touching the bar when you lift it.

Tips

This method supports the comb from below. The fabric strips will act as a sling, supporting the weight of the comb and holding it aligned with the top bar. The weight of the comb can cause wires or string that have been used to "sew" the comb to the bar from above to cut through and the comb to drop off. This is especially true if the comb is fresh and soft.

Space the strips for each bar as necessary to provide the best support; this will be based on the size and shape of each individual comb.

Avoid using bulky things (such as plastic hair clips) that force the top bars apart, as this will change the all-important spacing of the combs in the brood nest. A working top bar hive needs to have all its bars touching each other, with no gaps in between, and no way for the bees to access the space above the top bars.

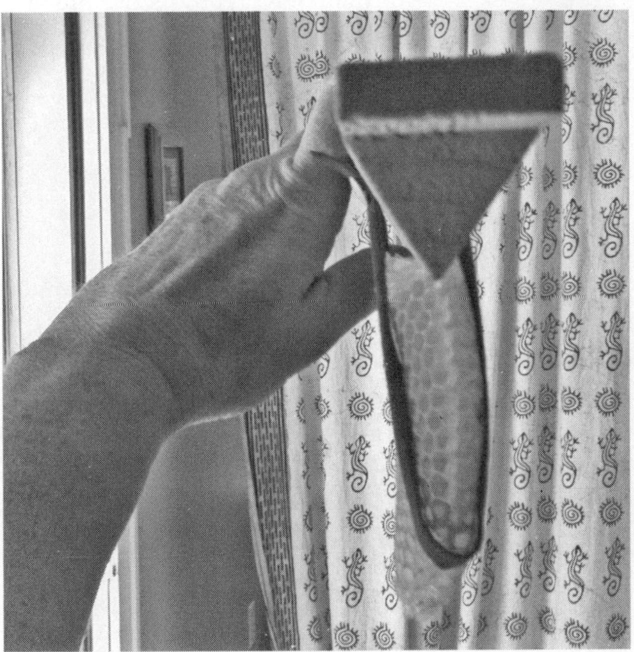

FIGURE 1.10. The sling method of rehanging broken comb. Credit: Christy Hemenway.

This sling method can also be used to correct a *cross comb* problem, or to reattach a comb that has collapsed off the bar due to heat or other factors. *Comb collapse* can best be avoided by giving your bees nothing but a dry wooden top bar to build from. The bees do the best job of attaching their comb to the bars that way, without any other input. Applying wax to the top bars is often linked to a comb collapse.

If your hive overwintered and is thriving in April, you have very different concerns than you did last year. Your next steps include a food check and a brood check, to be confident that things are on track. Then what? Now you should be thinking: Swarm alert! It's become more common in recent decades for colonies started from packages to *swarm* during their first year. It is even more likely—probable, in fact, and important—that a colony will swarm in their second year.

So let's move on—we'll cover the details of your second-year bees and their reproductive process in Chapter 2.

Swarming and Splitting *or* If You've Still Got Bees, You Are About to Have Even More Bees!

2

WELCOME TO BEE SEASON #2. If you've made it through April, you're officially into your second year with your top bar hive. So what are your bees going to get up to now? They already have comb built from last year, there's brood in the hive as well as honey, it's warm out, and there's plenty of *forage*. The next thing on your bees' minds is...the making of more bees. So let's talk about swarming and splitting, and let's start with a definition.

> swarm, *noun* \'swȯrm\ *a*: A great number of honeybees emigrating together from a hive in company with a queen to start a new colony elsewhere; *b*: A colony of honeybees settled in a hive

Most importantly, swarming is the reproductive mechanism of honeybees; and secondly, it is one of the bees' primary motivations, the driving force behind much of what they do.

At some point, Hollywood got hold of the word swarm, and they made it into a scary thing for the movies. Beekeepers who depend upon honey for their living don't like the fact that their honey-making forces get divided roughly in half. The incredible energy generated when a swarm departs its original hive can be both awe-inspiring and a little

FIGURE 2.1. This swarm was like a gift—very close to the ground!
Credit: Christy Hemenway.

frightening. In any case, whether you were hoping it would happen or hoping it wouldn't, a swarm is a special event.

I think of it as a "Congratulations, I'm sorry" event. Why? Because it *is* the reproductive mechanism of honeybees: it's when one colony of bees decides it is big enough and bad enough to turn itself into two colonies. It also means that your bees were strong and healthy and successful enough to make more bees. To that one can only say, "Congratulations!" There's no need to think of swarming as a loss of your bees because, in reality, it means more bees.

It does mean that the population of the original hive is reduced, divided roughly in half. To that, one could say, "I'm sorry." (I wouldn't say that, but one could.) The point I'm making is that swarming is a very important part of the bees' natural life cycle, and they need us to support that system rather than attempt to thwart it.

So know this: If you're coming out of winter and you've got bees, you are about to have *more* bees. (Congratulations!) But, since so many top bar hives are urban backyard beehives, it's important to know and understand the swarming process and what it means for you and your neighbors should your hive swarm.

Signs of Swarm Preparation by the Bees

When your bees are preparing to swarm in order to reproduce, you will see the following signs of their preparations in your early season inspections.

Workers

A hive prepared to swarm will have built up a large, strong population of female bees. The female bees are called *workers,* and they've been working hard. Many of your top bars will have had comb filled almost entirely with worker brood. What a beautiful sight! (Congratulations again!) The cappings of worker brood in the pupal stage are very slightly raised but flat and fibrous in appearance.

Drones

Another healthy sign of an imminent swarm is the raising of *drones,* the male bees. You will see drone-sized brood cells being *drawn out.* These

FIGURE 2.2. A worker bee (*top*) and a drone (*below*) face off at the hive entrance. Credit: PM Oakley.

FIGURE 2.3. The eyes of drones are huge—they cover their whole face, and touch at the top! Credit: Ben Sweetser.

cells are larger than the worker-sized cells. When the bees draw their own natural wax, without the use of foundation, they are more easily able to do this when and where they need drones. Sometimes the bees reuse existing drone cells, sometimes they remake existing comb into drone cells, and sometimes they draw out drone cells in brand new wax. They are usually located in clusters scattered among worker brood cells, often at the edges of comb. In the pupal stage, the cappings of drone bees are very distinctive. Unlike the flat worker bee cappings, drone cappings are dome-shaped. They look a bit like crunchy puffed cereal. (Yay! Drones!)

Queen Cells

A queen bee is raised in a very different cell from the cells of worker bees and drone bees. Worker and drone brood are both raised in cells that, when the comb hangs down as it naturally would in the hive, are oriented horizontally. But a queen cell must be oriented vertically. Swarm queen

FIGURE 2.4. Classic swarm cells on the edge of natural comb. The top one has hatched.
Credit: Jim Fowler.

cells are built on the edges of the comb. An advantage of natural wax comb drawn from a top bar with no wooden frame surround is that the edges of the comb are readily accessible to the bees. This is important for the making of queen cells. The wooden frame of a Langstroth hive gets in the way of the bees making these cells; you will see them struggling to make *queen cells* drawn awkwardly over the edges of the wooden frame. (And once again, congratulations are in order here!)

Queen Cups

A *queen cup* is the beginning of a queen cell. Rounder, shallower and less developed than a full-blown queen cell, it looks almost like they are practicing, or just "getting ready to get ready." Sometimes it's hard to tell whether the bees are completely serious about swarming because they may make and then destroy numerous queen cups throughout the season. When the bees have committed to swarming, the queen will lay an egg in the queen cell, even in what may have been just a queen cup, and very soon the bees will draw out the cup, to complete a full queen cell surrounding that egg.

FIGURE 2.5. A worker bee investigates a queen cup. Credit: Ben Sweetser.

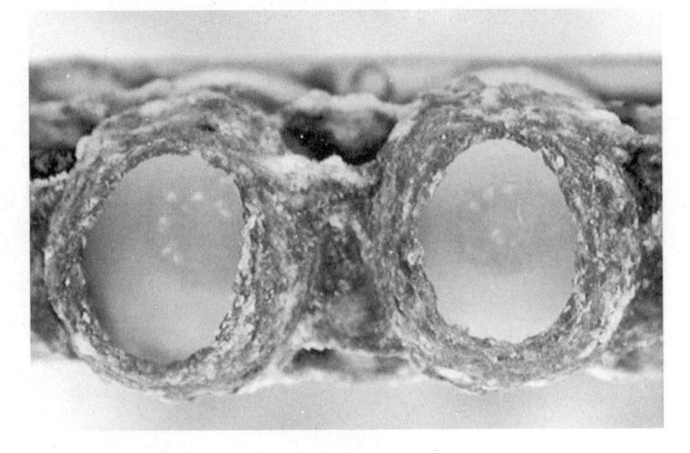

FIGURE 2.6. Queen cells with larvae. Credit: Christy Hemenway.

Nectar-packed Brood Nest

Probably one of the most confusing indications of imminent swarming is the backfilling of the brood nest with nectar. In conjunction with this, you will see the amount of new brood drop significantly, as nectar now fills the cells the queen would otherwise lay eggs in. This leaves a lot of nurse bees with nothing to do; at the same time, it reduces the size of the brood nest. The brood nest must become small enough that it can be cared for by the population of the *remainder hive,* in order for the hive to swarm. This reduction in size can take a full brood cycle. When you see the brood nest is filling with nectar, know that the hive's attention will soon be focused on making queen cells for swarming. A nectar-packed brood nest gives you advance notice of a reproductive swarm.

When the brood nest is packed full of nectar, the queen is unable to find a place to lay her eggs. She is spending a great deal of her time running all around the hive, looking for empty cells but finding none. This

FIGURE 2.7. A brood nest filled with nectar prevents the queen from laying and is an early indicator of swarming!
Credit: Dave Freytag.

frantic search causes her to lose weight, making her slimmer and better able to fly, while also making it more difficult to spot her.

Preventing Swarming

Preventing swarming at any cost is *not* the goal. There have historically been such practices as clipping the queen's wing(s) or destroying queen cells when the bees make them. These generally end in dismal failure. Clipping a queen is an attempt to prevent her from flying; the logic being that if the queen cannot fly with the swarm, then the bees will not swarm. This is a fallacy. It is typical that the existing mated queen flies with the swarm. But if the existing queen cannot fly, there is nothing to prevent the swarm from changing the timing, and departing with one of the newly hatched virgin queens. Do not destroy the queen cells you see your bees making. If you do this, and the bees swarm anyway, which is likely, you will simply leave the remainder hive *queen-less.*

Instead, think ahead! Prepare to have more bees by working with their natural behavior, but preempt the actual swarm by performing a split. Do regular inspections 7 to 10 days apart early in the season. Remember that two weeks is all it takes for a hive to hatch a new queen. Remember, too, that managing the swarm impulse appropriately will require having additional hives on hand, or aspiring beekeepers waiting in the wings. Do your best to stay ahead of your bees' needs.

Slowing the Swarm Impulse May Stop It

Bees choose the timing of the swarm in response to a complex combination of factors. The primary reason for bees to swarm is to reproduce, to create another thriving colony of bees. One factor that helps bees develop the confidence to swarm is that their hive cavity has become full. This lends credence to their feeling that they are large and robust. So being crowded, even if they are a small colony, can factor into the swarming process. The top bar beekeeper can work with this behavior by increasing the space the bees are occupying or by expanding the brood nest, thus reducing the bees' sense that they have filled their space.

Increasing Their Space

If your bees have filled all the space between the two follower boards or between the hive end and a single follower board, but your top bar hive still has empty bars, then it is easy to increase the space the colony is occupying. You simply add more blank top bars into the newer end of their brood nest, moving the follower board towards the end of the hive.

If every bar in their space has comb built on it, then they definitely need more space! Having 2 or 3 (at most 4) open bars available at the new end of a hive that is actively building up will give them more room to grow. Be conservative here, since more than 4 bars available encourages them to cross-comb, locking the bars together. Cross-combing quickly becomes a disaster in a top bar hive, making it impossible to inspect it.

Opening the Brood Nest

Adding space directly into the brood nest is known as *opening the brood nest*. Add 2 or 3 blank bars into the newer end of the brood nest, alternating the empty bars with existing combs. The size of the colony will help you determine how many bars to add. This encourages the bees to go back to work building wax to draw comb in the empty space, and it gives the queen more space to lay eggs. This helps to create the sense that they are not so large as they might have thought. If you catch them before they start to build queen cells, this technique can effectively work to slow them down; it may even stop them. If need be, you can repeat this process again as the season progresses. Adding blank bars in this way is somewhat disruptive to the colony, since the bees build each comb to fit specifically against the comb next to it, so be conservative here as well. Add only one blank bar between two existing combs, and don't add more than three blank bars each time you do this expansion.

Preempting the Inevitable Swarm

Another way of managing the swarm impulse is to *preempt* it—to allow the swarming process to progress naturally, but catch it before the swarm actually flies, and to purposely decrease the number of bees in the hive by dividing the hive.

Preempting a swarm is responsible urban beekeeper behavior. Frightening your next-door neighbors doesn't make for a happy or safe environment for you or your bees. Letting your bees swarm may also mean that they choose a new home that does not welcome them. A swarm moving into the wall or the roofline of your neighbor's house is rarely a desirable event.

If you live in a densely packed urban or suburban area, the sensation a swarm can cause is a very good reason to take timely, thoughtful steps to work in concert with this natural process. Or maybe you want to intentionally increase the size of your apiary. That's another good reason. Or perhaps you are actively seeking to *pay it forward* by helping another beekeeper get started with bees. That's a great reason too! In any case, by understanding what your bees are doing, you can support and protect their natural reproductive process and fulfill some of your own goals as well.

How to Split Your Top Bar Hive

The best way to preempt a swarm is to *split* the hive in question. Let's say you inspected your hive today and saw that it is chock full of bees, and there is plenty of brood in all stages: eggs, larvae and pupae. You may or may not be seeing a few dry queen cups; you may or may not see some drone brood. It's likely this hive is in its second (or third or even more) spring, though it is possible that it could be a flourishing first-year hive. The season is early, the weather is warm, and the nectar is flowing. Congratulations! You are about to have more bees—why not have them on purpose?

Or... maybe today's inspection showed multiple queen cells along the edges of the combs that are fully drawn, or capped, and a burgeoning population of bees, along with a brood nest full of nectar where there should be baby bees. Congratulations! You have successfully procrastinated to

Interchangeable Equipment

I cannot emphasize enough how important it is at this point to have *interchangeable hive equipment.* Splitting hives requires that the top bars and the comb of the source hive will fit into the hive body of the receiving hive. This will make the split a simple process—easy on you and easy on your bees.

If the top bars that your booming hive has been building on won't fit into the hive you want to move the split into, you will be out of luck, or you will find yourself hacking and slashing brood-filled comb to make things fit.

FIGURE 2.8. Interchangeable equipment is important! The Gold Star top bar will fit all top bar hives having an upper hive body width of 15 to 16 inches. Credit: Christy Hemenway.

the point where you are about to have more bees by accident! (What? You weren't ready? Ack!) The swarm may be a bit of an accident on your part, but you can still work with the process.

Whatever your situation, finding yourself with more bees can be either a very welcome thing or a very scary thing. It is helpful if you are knowledgeable about the bees' intentions, informed about the natural process of swarming and in tune with your own needs as a beekeeper.

Splitting a hive means just that. Begin with a thriving hive, with plenty of brood and bees, and divide it. Usually it divides in two, but sometimes, with an especially booming hive, you may be able to create three or even four new colonies.

Have on hand the additional equipment you will need. This means an empty top bar hive that is compatible with your existing equipment. Since you are preparing in advance, you have likely already made room for expansion, with a site chosen and the hive ready and waiting. If not, you will need to make arrangements for this as well.

It is often the preparedness of the beekeeper that determines the type of split that gets done. Interestingly, the easiest of splits is also the earliest and the most intentional.

The Walkaway Split—for the Prepared Beekeeper

At this stage, with a full and thriving hive that is showing no signs of imminent swarming, you are a prime candidate for splitting your hive in the simplest way there is, a walkaway split. We call it that because you can simply split them, and then walk away. The bees can take care of things from there.

A walkaway split is an early split. This beekeeper is working toward the goal of increasing the number of hives in their own apiary, or providing bees for other beekeepers. A walkaway split requires that the source hive be sizable and thriving fairly early in the season, and it takes intentional and proactive action by the beekeeper. It needs to be done early in order to happen before the bees decide to swarm. This beekeeper must be able to recognize that the hive is strong enough for splitting, yet has not begun swarm preparations on its own. For this reason, it is likely to be a second-year hive or older. Older hives that have overwintered will be larger and better established earlier in the season than most first-year hives can manage, especially those started from package bees.

A walkaway split is an easy split because it does not require searching through the hive to find the queen, or capturing her in order to relocate her to one hive or the other. Catching and/or moving a queen can be risky: she may get injured or killed. Since the bees are not yet intent upon swarming when this split occurs, there are no queen cells, so there are no worries about how you divide up the brood bars, and no need to worry about which hive contains the existing queen.

Why It Works

It is the bees' natural ability to create a queen bee from a worker bee larva that makes this split work. This allows the beekeeper to perform this split quickly and easily, and without having to know where the original queen wound up.

There is one very important requirement when doing a walkaway split: Both colonies *absolutely must* have the resources it takes to raise a new queen, i.e., combs containing three-day-old larvae. Without them, the bees will be unable to raise a new queen, and you will soon find yourself with a laying worker in a queen-less colony.

How to Do It

Inspect the hive with an eye toward taking an inventory of its contents: How many total bars of comb are there in the hive? How many combs contain open brood in the larval stage? How many bars have capped pupae soon to emerge? How many bars of honey? Are you certain that there are no queens already started? Be sure to examine the edges of the combs, where swarm cells are typically found. If you find unanticipated swarm cells, you will need to change your technique for splitting to what we call the Procrastinator's Split, as described next.

Knowing what you have to work with, you can now carefully divide the source hive by moving half the brood and food to the split hive and leaving half in the source hive. Do what you can to maintain the order in which the combs were originally built. Be aware that, on any given bar, you may be moving your queen, so be very cautious when handling the combs. It's a good practice to have a *nuc box* on hand when performing any split, to carry the combs you plan to move to the new hive. That way you are not carrying unprotected bars of comb across your bee yard, potentially dropping the queen and killing her by stepping on her.

FIGURE 2.9. A nuc box is super useful for moving or storing combs. Credit: Christy Hemenway.

Since it is difficult to accurately gauge the age of a young larva, and since it is so important that both colonies have what they need to make a replacement queen, another good practice when inventorying and dividing resources is to be sure that both hives have combs that contain open brood in all stages, both eggs and larvae.

Availability of Drones or… When the Boys Are

Any time a colony needs to make a queen, she must fly to mate. At the time of the split, drones must be available to mate with the new queen. This means that splitting is tied to your local season. Pay attention to what your own hive is doing; if you don't see any drones being raised in your hives yet, then it's too early to do a split!

The Procrastinator's Split or "Oops! I Wasn't Ready for That!"

When we call a beekeeper a procrastinator, we mean that in the nicest way, of course. What with work schedules, weather, vacation time and all the things that need doing in the summertime, the bees sometimes get ahead of the beekeeper. But if your bees have already begun preparations to swarm, you need to handle things accordingly. The timing makes all the difference.

So you've finally "gotten around to" doing a full bar-by-bar inspection of your hive. It's been a few weeks, and frankly, they look great! You have many combs, covered with many bees, and you see plenty of eggs, larvae and capped worker brood. You also see some drones and drone brood. On the edges of some combs, you see…"Oh, no! Are those queen cells?"…It's likely that they are indeed queen cells, and these cells may already be capped, or they may contain a developing queen larva resting in a shiny pool of royal jelly. If the cells are just tiny cups, and "dry," meaning they don't contain a larva, you may still be able to slow the swarm impulse by opening the brood nest, as described earlier.

But queen cells, on the edge of the combs, capped or containing a larva are signs that say a swarm is definitely on its way. Congratulations!

So now what? Well, a split is still a good plan, but now it's a little more complicated. One of the challenges of this situation is that the beekeeper who is most likely in this predicament is probably a relatively

FIGURE 2.10. This colony has made many queen cells—they are really hedging their bets!
Credit: Internet poster.

new beekeeper, one who is still learning to spot the queen and recognize the signs of swarming, and here they are with a swarm about to happen. This beekeeper often has no empty equipment on hand either, so where will the new colony go? It's always challenging when you find that you have more bees than you have boxes.

This impending swarm may also set up a conflict for the beekeeper: what if you didn't want more hives? Perhaps you don't have room, or perhaps you don't have time, or perhaps your town has an ordinance governing how many hives are permitted in your location. You are about to have more bees, but they need a new home! This situation is ideal for helping to launch another top bar hive beekeeper in a very supportive way. Between the two of you, the split can be performed, and the new colony moved to the new beekeeper's apiary—helping someone else get started with bees raised on natural wax, while relieving you of the need to manage more hives when your own apiary is already as full as you can stand.

Why It Works

The events unfolding in the source hive are what the bees would do naturally. They have built up to the point that they are confident that they have what it takes to turn themselves into two hives. The swarm impulse is strong and important, and even if the beekeeper wasn't prepared for it, the bees certainly are. They've created everything they need to reproduce: brood, food, and queen cells. It is the beekeeper's task to divide those resources appropriately. In addition, this split requires that the beekeeper be able to locate and move the queen. If the queen and all the queen cells wind up in the same hive, the other hive will suddenly find they need to start over making a new queen when they thought they had several in the works. It's likely this will cause a worker bee to begin laying eggs, this is known as having a "laying worker," a difficult situation to correct.

How to Do It

Begin like you would any other split—with an inventory of the contents of the hive. Pay attention to the total bars of comb, how many bars have brood (both open and capped), as well as honey and pollen stores. Make a special note of the bars that have queen cells on their combs. You can mark these by writing on the top bar with a pen, or sticking a pushpin into the wood. All the bars with queen cells *must* be housed together in the same hive after performing the split. This will affect the bars of brood and food that can be moved.

While you are going carefully through the hive and making notes, be on the lookout for the queen bee.

FIGURE 2.11. **Queen in queen catcher.** Credit: Christy Hemenway.

FIGURE 2.12. This colony is preparing to depart as a swarm. Don't confuse this activity with *bearding*. Credit: Dave Freytag.

The safest way to perform this split, which means moving lots of bars around as well as finding and moving the queen, is to find the queen and safely contain her in a *queen catcher* or *queen cage* until the chaos is over. Then you can calmly release her back into the hive. Using a queen catcher is a good way to protect your queen. You simply squeeze its handles to open it wide, maneuver it to where the queen walks into it and carefully let it close, trapping her safely inside.

Divide the hive's resources as nearly in half as you can, allowing for the requirement to keep all the swarm cells in one hive. Carefully move the bars you've earmarked to go into the new hive by putting them directly into that hive, or by using a nuc box to hold them while they are being transported to their new home. Again, do your best to respect the original order of the combs in each case, as the bees build their comb according to where they are located and what they will be used for, so the less disruption to that order, the better.

The Honeymaker or If/Then Split— "IF They Get Serious About Swarming, Then I'll Split Them."

This beekeeper would really rather that their bees made more honey than make more bees. Bees who are not devoting their energies to swarming have more bees available to focus on gathering nectar and turning it into honey. This beekeeper will also be well-versed in identifying the appearance and behaviors that indicate swarming. They've inserted blank bars into the brood nest to open it up and slow down the swarm impulse, as discussed earlier. If the bees prove uncooperative about staying home to make honey, then the beekeeper can be a bit philosophical about it, choosing to shrug their shoulders and work with supporting the bees' swarm impulse.

Why It Works

It works for the same reasons the Procrastinator's Split works: The events unfolding in the source hive are what the bees would do naturally. They build up to the point that they are confident that they have what it takes to turn themselves into two hives.

How to Do It

This split is much like the Procrastinator's Split, except that events have transpired to this point with the beekeeper's knowledge. This beekeeper is ready to split the hive because the bees have shown that they are truly committed to swarming. Otherwise the beekeeper would likely not have split them at all, in the interests of there being more surplus honey to harvest.

While you are going carefully through the hive and making notes, be on the lookout for the queen bee. Just as you would with the Procrastinator's Split, which means moving lots of bars between hives, your smartest bet is to find your queen and safely contain her in a *queen catcher* or *queen cage* until the split is complete, then release her back into the colony that has no queen cells. A queen catcher is a very easy way to protect your queen. You simply squeeze its handles to open it wide, maneuver it to where the queen walks into it and carefully let it close, trapping her safely inside.

Splitting into the Same Hive or How to Buy Yourself Some Time

In the event you have no aspiring beekeeper waiting to receive your bees and no extra empty hives, a Gold Star top bar hive gives you the ability to split the hive and house both colonies in the same hive box.

Why It Works

The Gold Star hive, which is a *side center entrance hive*, has three entrances located in the center of the long side, the front of the hive. It has a single entrance located at each end of the back of the hive. It includes two follower boards, which can be used to manage the hive body in different ways. These features make the hive very versatile should you need to use it to house two colonies.

How to Do It

Like you would for any of the splits we've discussed, inventory the contents of the hive. Divide resources equally, managing them as needed, based on whether there are queen cells or not. Place the two colonies in opposite ends of the hive. Then place the follower board(s) in the center, positioning them so that they allow access through the front entrances for one colony, but not for the other.

The colony that contains the queen cells should occupy the half of the hive that will continue to have access to the front doors, the doors they are accustomed to using, since this is the colony that would stay behind. The other colony, where the existing queen resides, will occupy the opposite end of the hive, and this colony will use the rear entrance at that end, simulating a move to a different location, as if they had swarmed naturally.

Be aware that this is only a temporary solution, since it limits the amount of space that the bees can expand into, but it can certainly buy you some time while you either acquire additional equipment or find another beekeeper with interchangeable equipment.

Pay specific attention to the weather and to the nectar flow while you have two colonies living in the same hive. In this configuration, there is no space to put a feeder for either colony. This can create a starvation problem should you find yourself in a serious nectar *dearth*. You may need to feed outside the hive to help prevent this, which increases the risk of *robbing*, or hurry to get one of the colonies moved into a hive box of its own.

Other Details Concerning Splits...

What About the Genetics?

After considering the logistics of splitting your hive, you're likely wondering, "Wait a minute—which hive gets which queen?" Good question. In a natural swarm, it is typically the existing mated queen that flies away with the swarm and begins the work of building up from scratch in the new location. The new queen hatches in the original/source hive, flies to mate and then returns to the source hive and begins laying eggs to repopulate it.

By this reasoning, the colony that contains the existing mated queen is the logical choice for moving to a new hive. The source hive will soon be a well-populated hive with a new queen, and would remain in the original location.

Consider this though—the original beekeeper may prefer to keep the original queen. After all, her genetics are known and likely desired. The new queen remains an unknown until she has successfully mated and returned to the hive, and begun laying eggs. How to counter this? Ideally, the beekeeper has the ability to move the original/source colony with the original queen—not far, but away from the original location—so that the bees have the sense of having flown. Otherwise there is some risk that the bees will swarm anyway, having already been involved in preparations to swarm.

If you are splitting into the same hive box, you use the same logic. The remaining half of the hive, which will contain the queen cells, will stay where they were, and will continue to use the original "front door" entrances, in the center of the long front side, and approximately one-half of the hive. The follower boards are placed in the center, positioned to allow access through the front of the hive by the bees that will remain in the old location. Place the follower boards so the bees that are using the front entrances have no direct access to the other end of the hive. The half of the colony that would have swarmed, the one that includes the existing mated queen, is housed on the other side of the follower boards, and they use the entrance on that end of the rear of the hive.

Where Did All the Foragers Go? Or…"Where's All the Activity?"

When splitting a hive, it's important to be aware of the final disposition of the field force, or foraging bees. After a hive is split, the field bees keep on doing what they had been doing—foraging—from where they had been doing it, the original or remainder hive.

Bees know their home as a particular "place in space," or GPS coordinates, if you will. They are not oriented to the appearance of the hive, but to its location. If you move the source hive, they will continue to return to the original location of the hive, even though it is no longer there. Because of this, the hive that contains the original queen, which you will move to a different location, will have very few foragers, as they will all join up with the hive in the original location.

Since the existing queen is still laying and brood is continually hatching, this imbalance will eventually sort itself out naturally, but the lack of activity can be nerve-wracking, especially if the hive is not heavy on food stores. There are some ways the beekeeper can help counteract this imbalance. The first, and simplest, is to move lots of bees into the hive containing the original queen. If the split is done in late afternoon or early evening, more of the bees will be in the hive, so this is easier to accomplish. Gently shake or brush bees from the combs into the original hive.

Another option is a bit like pulling a *bait and switch* on the bees. After the split has been done, give the bees a day or two to settle and reorient to their new location. Then switch them. Move the original hive back into the original location and let it collect some of the field force. A day later, move each hive back to its desired location.

The Integrated Pest Management Aspect of Swarming

So why do we put all this emphasis on being in favor of swarming, and splitting, and increase and managing more bees? One reason is that swarming naturally causes a *break in the brood cycle.*

A lack of brood is a serious setback for the *varroa mite,* which requires the open brood cells of the honeybee occupied by larvae in order to start their reproductive cycle. The adult female mite begins this life-cycle phase by sneaking down inside an open brood cell and hiding beneath the larva floating there in a shiny little puddle of royal jelly. She waits for the cell to be capped, then begins laying her eggs.

In the remainder colony—the one that stays behind and must make a queen—there are no eggs being laid, and the existing brood is hatching out daily. The new queen must hatch, fly to mate, successfully return to

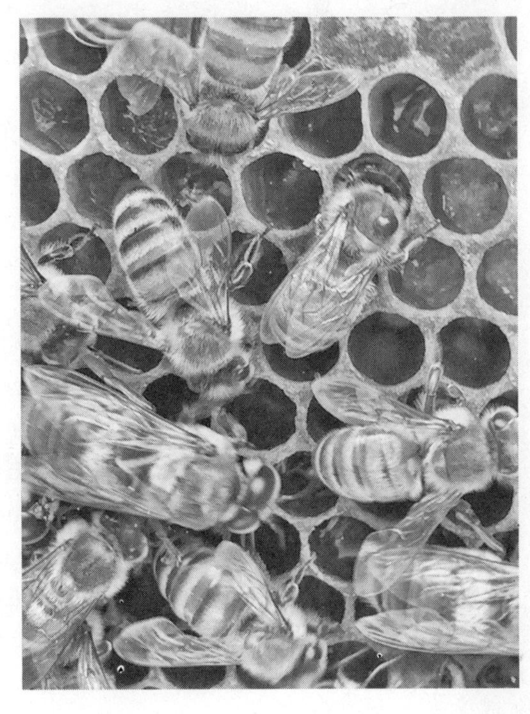

FIGURE 2.13. (*above and right*) Varroa are ubiquitous. Occasionally you will see them on the backs of the bees. Credit: Ben Sweetser.

FIGURE 2.14. A "baton" swarm. Credit: Christy Hemenway.

the hive and begin laying eggs. This takes approximately 28 days. By that time, all the brood in the hive will have hatched, and there will no longer be open brood available in the hive for the female mite to lay eggs in.

In a natural swarm, the colony that flies with the mated queen leaves behind all their existing brood and must begin building their hive from scratch. So for the colony that flies, there is no brood for varroa to breed in.

The less brood there is in the hive, the fewer cells available for the varroa mite to breed. This is therefore a good way to control varroa without using chemical treatments. Dare I say it again...? Congratulations!

The swarm impulse is strong and biologically important to healthy bees. Managing swarm behavior by working proactively with the bees to preempt the process before they actually fly retains the advantages of natural swarming and avoids most of the disadvantages.

How a Beekeeper-Induced Split Differs from a Natural Swarm

When the beekeeper splits their hive, the colony that would naturally fly to a new location and begin building their hive from scratch does not get the opportunity to do that. The queen of that colony continues laying eggs in the combs that the beekeeper moved with that colony, so there is no break in the brood cycle for this colony. One method of compensating for this disadvantage is to purposely create a break in the brood cycle by *caging the queen*. This is detailed in Chapter 3 and 12.

Overwintered = Survivor Stock

This new queen is the offspring of the older, proven queen. If the hive that is being split is a second-year hive, then both queens and both colonies now stand a good chance of overwintering again, helping to propagate locally raised *survivor stock* in your local area. Congratulations!

Hurrying a Walkaway Split

The time it takes for the bees to replace their queen can be alarmingly long, especially in a climate with a very short growing season. Artificially replacing the queen can decrease the time it takes for a colony to recoup after a split, bridging a timing gap that might otherwise spell disaster for the hive as the local foraging season comes to an end.

Unlike the timing of the natural queen replacement that occurs with swarming, or with preemptive splits that support the bees making their own queen, the beekeeper performs this manipulation. It is tied to reproduction and the multiplication of hives, but bypasses the natural break in the brood cycle that occurs when the queen is replaced naturally by the bees.

Why It Works

Bees move the queen's pheromones throughout the hive by touching each other as they move within the hive. The level of queen pheromone in the hive dissipates very quickly if she is removed. With the loss of the queen and the subsequent drop in pheromone level, the bees rapidly become aware that they are without a queen and go into queen replacement mode within hours.

When the hive realizes they are queen-less they also become more receptive to a new queen and are therefore less likely to reject and kill a new queen, if one is provided.

How to Do It

Prepare to perform the split. Have a mated, caged replacement queen in hand to prevent any unforeseen circumstances in obtaining her. Perform the split as previously described, and then wait 24 hours before installing the new queen. Utilize the slow-release mechanism of the queen cage to allow the bees to adapt and accept the pheromone of the new queen. If she has not been released from the queen cage in three days, hold the cage deep down inside the hive and remove the screen, releasing her directly into the hive.

What If I Missed the Signs of Swarming?

Drones, queen cups, queen cells, nectar in the brood nest, a crowded hive...all these things taken together are indicators that you will soon have a swarm of bees issuing from your hive. If you didn't see it coming, or take preemptive action to intercept them, that swarm will soon be flying off to start a new colony in a location of their choosing. What a sight to see that will be! Thousands of bees will take to the air all at once, in a large funnel cloud, making a phenomenal sound—sometimes described as a low-flying airplane, sometimes as a distant freight train, sometimes as a nearby blow-dryer. It's amazing the amount of noise these tiny insects can make.

FIGURE 2.15. A swarm in the air. Credit: Gold Star customer.

They will fly as a group to an interim *bivouac* location and gather there in a tight cluster—a ball of bees. This is when bees and beekeepers often find themselves in the news. To see this many bees together, acting in concert outside their hive, is admittedly a bit sensational. With the increased public awareness of bees in these troubled bee times, it gets a lot of attention. You've probably seen news stories of bees swarming on cars, mailboxes, tree branches, storefront windows, bicycle kickstands… just to name a few. Maybe a swarm is the reason you now find yourself keeping bees!

The ironic thing about the drama caused by a swarm of bees is that while a swarming colony of honeybees looks amazing, it is when the bees are swarming that they are most docile. One reason for this is that, before they flew from the hive, they had filled their bellies to the brim by gorging on their honey stores, and so they are a bit logy on this flight. Another reason is that they are in a very fragile situation, striving to complete this part of the swarming process successfully. They are on their way to starting a new home in a new cavity. They are not protected by a physical structure; they are not out foraging to gather more nectar or pollen; they are simply clustered tightly together in a special sort of way—and they are quite still, almost dormant. Another reason for their docility is that they have none of the precious resources that bees fight hardest to protect. When hanging in a ball from a tree branch, they have nothing—no brood and no food—to defend. So, aside from their need to stay together and be accompanied by their queen, not much is of interest to them, least of all humans.

While all this swarm behavior is happening, 300 to 500 scout bees will have fanned out in the local area, searching for an appropriate cavity for the swarm to move into. According to Dr. Tom Seeley in *Honeybee Democracy*, an appropriate cavity would be about 40 liters in volume, have an entrance of approximately 3 square inches and be located 15 feet or more above the ground. The dynamics of the decision-making process that is occurring at this point in the hive's life cycle are beautifully detailed in Dr. Seeley's book, based on his years of researching swarms and their behavior. Through a process of *dancing* on the surface of the hive—communicating information about the size, location and suitability of the possible new location—the bees choose which cavity the scout bees have located that they will move into. Having decided, the swarm will "rev up" and fly off en masse to their new home; this happens very quickly, taking only a couple of minutes. Once there they will begin again to create a thriving colony of honeybees.

Catching a Swarm

So...your bees have swarmed. Or someone else's bees have swarmed, and you got the phone call because everybody knows you're keeping bees because you've been talking about it so much. The bees are outside the hive and hanging in a ball from a branch or a fence rail, or under the eaves of your house or wherever they chose to stop on their way to their new digs. So there's a swarm on the loose. Now what do *you* do?

Hopefully you have been preparing for this moment since you got your first bees, knowing that bees are striving to do just what they have done—swarm—to make more bees. If you have, then (Congratulations!) you very likely prepared by having an empty hive available, waiting for these new bees.

Now your challenge is to interrupt their home search process and introduce them into the new home of *your* choosing. This isn't complicated, but it does need to happen quickly. Bees may only stay in their swarm cluster for a short time (sometimes minutes, often only a few hours), while making the official choice of new location. Knowing this, many beekeepers that are interested in collecting swarms of bees keep a "swarm kit" at the ready. This allows them to respond quickly to swarm calls from concerned citizens, emergency personnel, pest control companies and others who need a prepared and knowledgeable beekeeper to collect some bees.

The contents of a mobile swarm kit might include:

- A bee-tight container. Often this is a cardboard box, but it could be a bucket or any other container with a lid. It does not need to be

airtight; in fact, you may want to add some ventilation by taping a piece of screen securely over a hole in the side of the container.
- Masking tape
- Your bee veil
- Step ladder (Caution is advisable when the swarm's location requires a ladder. Even though a swarm is generally very docile, being startled by the sting of an accidentally crushed bee may cause you to lose your balance and fall. A broken leg can make beekeeping difficult, so "bee" careful!)
- Lopping shears
- Small hand trimmers
- Extendable tree trimming saw, especially one with a hook on the end
- Sheet or tarp
- GPS or map or directions to the location of the swarm
- Cell phone
- Camera
- Piece of cardboard
- Bridge for walking swarms found on the ground into your container

Capturing the swarm is usually fairly simple. The swarm cluster is made up of bees hanging on tightly to other bees. Only those that are actually clinging to the surface that the cluster is hanging from are holding onto anything besides other bees. You have only to hold the box or other container below the swarm cluster and jar them loose. That's pretty easy to do in the case of a branch. Use your fist to strike the branch and the force of that impact is usually enough to cause the entire ball of bees to fall into the box. Be ready for the weight of the bees when they fall; a swarm can weigh several pounds, and the shock of their weight can cause you to lose your grip on the box if you aren't prepared for it.

Sometimes a swarm is wrapped around a "y" in a tree (gnarly old apple trees are notorious for this, it seems), and no amount of you jarring or knocking the branches is going to dislodge them. In such cases, having a piece of cardboard that you can use to gently scrape along the branch,

FIGURE 2.16. Collecting a swarm can mean climbing some very tall ladders!
Credit: Gold Star customer.

FIGURE 2.17. Collecting a swarm into a "high-tech swarm collector"—a cardboard box.
Credit: Gold Star customer.

FIGURE 2.18. Swarms that fall to the ground will "march" over a makeshift bridge into a swarm box. Credit: Christy Hemenway.

causing clumps of bees to fall into the container you are holding below, is very helpful. If the property owner is not dismayed by you cutting or trimming the branches of the tree, this can simplify the process.

Placing a tarp or sheet on the ground beneath any swarm can be helpful if part of the cluster winds up falling. Getting bees out of grass is difficult, so if they drop onto a sheet you can simply use it to gently gather them up and empty them into your container. If you encounter a swarm of bees on the ground, be sure to look up—above those bees—because you may find that the ones on the ground fell from a swarm cluster that is hanging above.

One method of getting bees out of the grass is to create a *bridge*. Many materials can work for this purpose. I've had success with the piece of white plastic used as a mite counting board in Gold Star hives, but a simple stick can also work. Place one end of your bridge in the cluster of bees on the ground, the other end at or near the entrance to your box. It's uncanny the way the bees will "march" across the bridge into the box, saving you the difficult task of trying to pick them out of the grass.

FIGURE 2.19. **You can use the flap of a box to gently nudge bees into your swarm collector.** Credit: Christy Hemenway.

Swarms that cluster on a vertical surface, such as a building or a vehicle, can be gathered by using a flap of the lid of your swarm box. Gently wiggle it in between the bees and the surface.

Once the cluster of bees is in your box or other container, you must be certain that the queen is amongst them. If you are fortunate enough to see her there, that answers that question, but spotting a queen in a mass of bees at the bottom of a box can be a bona fide challenge. However, the bees will tell you what you need to know. Spend some time observing, watching the bees that are inevitably in the air at this point. If the queen is inside the box, the stray bees will fly to the box and rejoin the cluster. If she is not, she is likely still at or near the original location of the swarm. In that case, the stray bees will begin to gather there again, and bees inside the box will leave to fly back there. That's your cue that you need to make a second attempt to gather the queen along with the rest of the cluster. Ten minutes of watching the bees will make it pretty clear.

When you are convinced the queen is in the box, along with as many of the flying bees that you have the patience to wait for, close up the container and secure its lid. For closing a cardboard box, I have found that masking tape is better than duct tape. It sticks better, which is definitely what you want. Having your swarm box begin to leak bees while you are transporting them in your car, or while it is sitting in your house as you prepare their new home, or while you are waiting for their new beekeeper to collect them is definitely not a good thing!

Installing the Swarm in Their New Hive

When you get the bees to their new home, installing them into the hive is fairly simple, especially compared to the detailed process of installing a package. Have the hive ready and waiting—lid removed, top bars out. *Bonk* the box firmly on the ground. This will cause the bees to drop to the bottom of the box. Hold the box over the hive and open the lid. Carefully pour and shake the bees into the hive. Since your queen is loose in this mass of bees, do your best to be certain that all the bees fall into the hive and not onto the ground. Stepping on the mated queen of a primary swarm would be a poor beginning! Put the top bars in place over the bees and put the hive lid on.

Recombining Hives or...No Dinks!

Sometimes a small colony of bees will swarm despite their size. This may leave the beekeeper with two small colonies, neither one robust enough to do well on its own. In this case, it makes sense to recombine the two by capturing the swarm if possible and simply installing it right back into the original hive, from whence it came.

For the beekeeper who wants to increase the number of hives in their apiary, this seems counter-intuitive, but it is a better plan to support one large thriving hive than to fuss and worry over two *dinks*, neither of which has a good chance of survival, especially through winter.

Swarm Trapping or...Another Way to Get Free Bees

If you are interested in increasing the size of your apiary, one way to go about that is to set up a *swarm trap*. This is a container such as a nuc box—a small version of your top bar hive—that can be hung in a tree, or placed in a location likely to be favored by a passing swarm. It should be equipped with several top bars that match your existing hive equipment. Locate the swarm trap where you can easily monitor it for activity, so that you will notice if a swarm moves into it.

Just as you do when starting a new top bar hive, you'll want to have a piece of brood comb inside your swarm trap to attract and to help anchor

FIGURE 2.20. Bees building bars of comb.
Credit: Dana Gray (*top*), A & R Verbeek (*bottom*).

the swarm. Either use bars that have comb attached or wire a piece of comb to one of the bars. Dribble a few drops of lemongrass essential oil on the interior wood of the swarm trap as an additional lure.

Be aware of two important factors regarding swarms caught in this way. First, a swarm is very organized; the bees in the swarm are all the right ages for all the jobs that need to be done when starting a brand-new hive. This means that there will be plenty of bees in the swarm that are of wax-building age. Swarms can build comb faster than you can say "Bzzzzzt!" I have seen a swarm build 22 full bars of comb in 17 days. In another case, they built 17 bars in 14 days. That's more than one full bar per day. So be ready! It's astonishing to watch, and the difference between a swarm and package will amaze you.

The second important factor is that the source of the swarm is likely to be unknown to you. This means the bees may be treatment-free, or they may not. They may be natural-sized bees, or they may have come from a hive using larger "standard-cell" foundation, meaning they are larger bees. Large-cell bees can size down, but they will first draw comb that is a bit bigger than the typical 4.9 mm that is considered the natural size of a honeybee brood cell. The vigor and organization of a swarm and their ability to build up quickly is a plus, but this may be offset by the bees' exposure to chemical treatments or the possibility of disease. Be aware of these things when installing swarms in an apiary already occupied by natural-size, treatment-free bees.

3 Why Are My Bees So Grouchy? *or* Is the Honeymoon Over?

I will arise and go now, and go to Innisfree,
And a small cabin build there, of clay and wattles made;
Nine bean-rows will I have there, a hive for the honeybee,
And live alone in the bee-loud glade.
And I shall have some peace there, for peace comes dropping slow,
Dropping from the veils of the morning to where the cricket sings;
There midnight's all a glimmer, and noon a purple glow,
And evening full of the linnet's wings.
I will arise and go now, for always night and day
I hear lake water lapping with low sounds by the shore;
While I stand on the roadway, or on the pavements grey,
I hear it in the deep heart's core.

— William Butler Yeats,
"The Lake Isle of Innisfree"

Early on, when a hive is just getting started, there is a sweet, romantic period during which the new beekeeper and their first bees are very much like a new couple on their honeymoon. The beekeeper is pleased to have successfully gotten started and is very much in love with their bees. The bees are sweet, gentle, peaceful, easy to be around and fun to watch. People marvel about how docile their bees are, about how they

FIGURE 3.1. Beautiful 3-pound package of honeybees.
Credit: Christy Hemenway.

never seem to need to wear any protective gear. They are convinced that their bees "know" them, and they assume that's the basis of the non-defensive behavior of their brand-new hive.

In reality, a new hive at the beginning of its first season starts out pretty small. A typical 3-pound package only contains about 10,000 bees, so there aren't that many bees in the hive yet, and they don't have anything but the queen to defend at that point. A second-year overwintered hive may be larger and have more at stake early in their second year, but compared to a colony at the height of the growing season, even a second-year hive starts out the season fairly small.

While it's nice to think of the beauty of "the bee-loud glade" that William Butler Yeats wrote about, the systems at work inside your top bar hive are not really about our human view of sweetness and light, or peacefulness and pastoral scenes. Those systems are about the bees' survival; they are striving to produce and protect more brood and more food.

It's important to remember that the bees' focus is on building up their numbers in order to get big enough to swarm. Queen bees are capable of laying 1,500 to 2,000 eggs *every single day*. By the time a new hive is six weeks old, this adds up to a *lot* more bees! At this point, the relationship of the beekeeper to their new bees is likely to change—maybe the new beekeeper even gets stung! The new beekeeper wails, "What did I do wrong? They were so nice before, and now they are so grouchy! They stung me!"

It's true. Bees sting. They don't do it to be mean, and they don't do it because they don't love you anymore. The worker honeybee stings to defend the two things that matter most to the hive and its survival: brood and food. So intent is she on the survival of that brood and food that she is willing to give her life protecting it. The sting of a honeybee is a one-time event: a kamikaze mission, a life-or-death moment. They don't sting for fun—and they don't seek you out purposely for the joy of stinging you unless you present a threat to brood or food. (Then all bets are off!)

So while it's likely that you will have a better experience if you inspect your bees calmly and gently, than you will if you crash roughly through your hive, you can't attribute either experience to your bees' knowing you, or even caring very much about you except for where you fall on their "perceived threat" scale.

It's difficult to prepare a new beekeeper for this shift in the hive's behavior, since it's difficult to describe the difference in size and energy between a brand-new beehive and an older, more established hive, and of course, everyone's perception is different and completely subjective.

Hostile Bees...or Wow, These Girls Are Really Mean...

So, the honeymoon is over, and your bees are all about the business of building up the colony now. Maybe they are a little more defensive during inspections, but nothing untoward.

But what if it's a bit more than that? What if every time you walk out into your bee yard, or your backyard, or every time you attempt an inspection or approach the hive, bees are hounding you? How do you discern whether or not you have grouchy bees, or some truly hostile *hot bees*? And if so, what do you do about it? Resolving this question may require some sleuthing. Some of the reasons that bees can become defensive include:

Environment

- **Nectar dearth:** Dearth is a fancy word for lack, or shortage or absence. At certain times in certain climes, in conjunction with weather conditions like high temperature and low rainfall, the nectar flow simply shuts off and becomes very scarce. This doesn't bode well for bees that are working to gather enough nectar to survive the biggest of dearth periods—winter—and can make them a bit testy.

- **Light:** Check for lights that are on during the night near your hives. Back-porch lights, spotlights, the ballfield next door, streetlights, motion detector lights... all these can irritate your bees.
- **Noise and/or vibration:** The constant sound and vibration of heavy equipment operating nearby can be a source of irritation. The noise of your lawnmower can sometimes alarm the hive. Bumping the hive with the lawnmower is almost certain to get a reaction.
- **Smells:** Bananas smell very much like alarm pheromone and have been known to send the bees off into a fit of hostility. Certain personal care products—shampoo, lotions and colognes—also seem to offend the bees. Do your best to smell as neutral as possible when you head out to inspect your hive.
- **Protective finishes:** There are many different ways to protect a top bar hive from the elements; some may involve substances that have a strong smell. If your choice of finish has a strong smell, or if it off-gasses, you may want to consider a different one. Be sure to allow plenty of drying time so that the smell of the curing finish is not irritating to your bees.
- **Nocturnal visitors:** Skunks are famous for scratching on the wood of the hive at night, to the point that bees go out to investigate the annoyance only to be eaten by said skunk. The scratch marks on the hive are your first indication of this intruder.

Hive Status

- **Queen-lessness**: Sometimes being without a queen makes the bees lethargic and listless, depressed even, but other times it makes them grouchy. It might be time to suit up and do a full bar-by-bar inspection, searching purposefully for the queen, or for queen cells, to see if that might be the cause.

- **Swarming:** Bees that are preparing to swarm can be more irritable than usual. Their whole lives are on the line, and they are in a very fragile situation. It makes sense that they might be a bit on edge, doesn't it?
- **Robbing:** A hive being robbed is likely to be irritable and defensive. See details below on robbing.

Genetics

- **Africanization:** *Africanized* honeybees are probably the basis of most scary Hollywood bee stories, and have gone far to create a fearful attitude regarding these bees. While they are not the gentle European honeybee most of us know and love, even Africanized honeybees can be worked with by knowledgeable and appropriately garbed beekeepers. However, hot bees—whether Africanized or not—in densely populated urban areas can present a very serious problem. Sometimes it's just in the genetics.

Re-queening

If you have checked for and eliminated any of the possible causes listed above, and checked in with yourself about your own level of experience, confidence, and comfort with bees, and still find that your bees are too defensive or potentially even unsafe to work with, the solution to consider is to *re-queen* your hive.

Re-queening has been used (sometimes perhaps a bit too casually) as a solution to everything that ails a beehive. Please be conservative about re-queening, but know that in the case of a very defensive hive, it is by far the fastest and most effective solution because of the short summer life span of the bees. The average worker honeybee lives for only about 42 days in the summer. That means that there are many bees dying and many bees being born every day during the growing season. This rapid turnaround creates an opportunity for quickly changing the genetics of the hive.

To re-queen your hive, first consider your options for a replacement queen. Be aware that replacing the queen will mean that the genetics of the entire hive will quickly become those of the new queen. If you chose your bees' genetics for specific traits, and the new queen comes from different genetics, which is likely, those traits will be affected by this change.

The speed of the change in genetics is the key to the success of re-queening as a solution to hot bees. The genetic changeover happens quickly; by the end of 42 days, all the worker bees in the hive will be the offspring of the new queen. In many cases, the temperament of the bees changes almost immediately, as soon as the queen is replaced.

Once you have the new queen in hand, or you are completely certain one is available, follow these steps to re-queen your hive.

For your safety, and for your confidence level, safeguard yourself well from your grouchy bees by wearing the appropriate protective gear. A veil is a must; gloves are necessary. Long sleeves and long pants are important too. Boots that protect your ankles are a wise choice, because bees can easily sting through your socks. You needn't purchase a full-length bee suit necessarily, but you might consider adding extra layers on your torso and legs, as it is possible to be stung through jeans and even the typical bee jacket or suit.

Begin to inspect your hive, looking specifically for the queen. When you find her, catch her and *pinch* her. Literally. (Yes, that is what I said, and yes, we hate it too.) Close the hive up and leave them queenless for a day. This is long enough for the queen pheromone in the hive to dissipate, and for them to realize that they are queenless. They will then be more receptive to the introduction of the new queen.

While you could direct-release her into the hive, it's probably a better idea in this

FIGURE 3.2. A queen cage with its *candy plug* acts as a slow-release mechanism. Credit: Geoff Keller.

case to utilize the *slow release* method. This will give the bees a little more exposure to her pheromone and more time to accept her.

Take a close look at the cage containing your new queen. There should be a plug made of sugar candy blocking one exit. The exit may also be plugged with a cork, or blocked in some other way in order to protect it from being consumed by bees outside the cage. Often, replacement queen cages are just that—the cage itself, with no method of suspending her in the hive. Choose a method of suspending the cage in the hive, knowing that in your top bar hive, your bars all need to remain touching. If there is nothing already attached to the queen cage, be resourceful. Attach a thin wire, or plastic strap or a thin strip of hardware cloth to the queen cage; something that will allow you to attach the cage so that it hangs from one of your top bars.

The next day, gear up again, and take the new queen, in her cage, out to the hive. Open the hive and choose a location to hang her. This can be on the end of a comb not fully built all the way across the bar, or on a top bar that is inserted into the brood nest or next to the brood nest. Attach the queen cage to the bar, and remove the plug or other obstacle protecting the sugar candy. Close up the hive. You may see results almost immediately. Go back in 3 days to ascertain that she has been successfully released, and now you can remove the queen cage. It's likely that your bees will be gentler at this point, so this inspection should be much calmer!

Robbing—When There's Not Enough to Go Around

Yes, it's true—by midsummer, your sweet and gentle honeybees can turn into thieves. Or, worse—someone else's formerly sweet and gentle honeybees can turn into thieves, and begin robbing *your* sweet and gentle bees. It's not pretty!

Either way, if robbing begins happening in your bee yard, it requires your *immediate* intervention. An entire hive full of honey can be emptied by robbing bees in a day, leaving it seriously decimated and devoid of food stores late in the season.

Still, you don't want to jump to inaccurate conclusions about robbing, if that is not what really happened. The tricky thing about identifying robbing is that your first clue is an unusual amount of activity at the hive; there will be lots and lots of bees flying. This is usually a cheerful sight, so the first thought to come to mind is, "Wow, look at 'em go. They are really busy today!" There are other reasons besides robbing for seeing increased activity at the entrance to a hive, some of which are not worrisome at all; so it's important to be able to tell the difference. Observing the details of the bees' behavior will help.

Things That Are Not Robbing

Orientation Flights

The most likely activity to be confused with robbing is the sight of young bees taking *orientation flights* in front of the hive. This *is* a cheerful sight.

What the Beekeeper Sees

In the midst of otherwise normal foraging activity, there will suddenly be many bees in the air outside the hive, all facing the entrance and gently hovering—up and down, side to side.

Orientation flights typically take place on sunny afternoons mid-season, when lots of brood is being raised. It is *not* so typical to see orientation flights later in the season when robbing is more likely.

The activity itself usually only lasts for 20 minutes or so. The appearance is calm, orderly and peaceful, and it looks as if somebody grabbed the volume knob and "turned up the bees."

Orienting bees will be young bees—they will look a bit "fuzzy"—as young bees still have most of the fine hairs that cover the thorax and abdomen. These fine hairs wear away with age, so older bees tend to appear more striped; younger bees are more solid colored, and fuzzier.

Drone Eviction

Another activity that can look like robbing is the eviction of drones in the fall. It's not necessarily a cheerful sight, but it is not robbing.

What the Beekeeper Sees

This also involves an unusual amount of activity in front of the hive. There will be bees wrestling as well; the worker bees take no prisoners when they are evicting the drone population as fall arrives. The sight of a tiny worker bee carrying a huge drone out of the hive and dropping him in the grass is pretty impressive.

In this case, being able to identify a drone is crucial, so that you can be sure what is happening. A drone is very bulky through the thorax and has a plump rounded abdomen, but the eyes are the most prominent feature. They are so large that they cover the front of the head, nearly touching in the center. Compare this to the smaller round eyes of a worker, placed well apart on the head.

FIGURE 3.3. The eviction of the drones occurs in the fall. Credit: Ben Sweetser.

Actual Robbing

Robbing can begin if a hive is small or weak. It's especially likely to occur during a nectar dearth. Sometimes a queen-less hive is so low on morale and energy that they also become easy targets for robbing.

What the Beekeeper Sees

What the beekeeper is likely to first notice when a hive is being robbed is the intense amount of activity at the hive. There will be many, many bees in the air, creating an appearance of tremendous industriousness. Upon closer look though, things are not so peaceful or so productive as they seem.

Robber bees are sneaky and nervous, erratic, almost frantic. You may see guard bees wrestling with robbers at the entrance, trying to protect the hive and its treasures. This battle will be to the death, and the toll that it can take on your hive's population can completely decimate the colony.

Also look for robbing bees that arrive at the hive carrying no nectar; they are flying light and quick. They may zigzag around looking for an opportunity to sneak past the guard bees. When they leave the hive, they are weighed down with their stolen goods, so now they are heavy and awkward. They may walk up the front of the hive before they depart. One particularly revealing clue is to watch them take off, then "sink" momentarily before actually gaining any altitude. This is in direct contrast to foraging bees, who are the opposite—light when they leave the hive and heavy when they return.

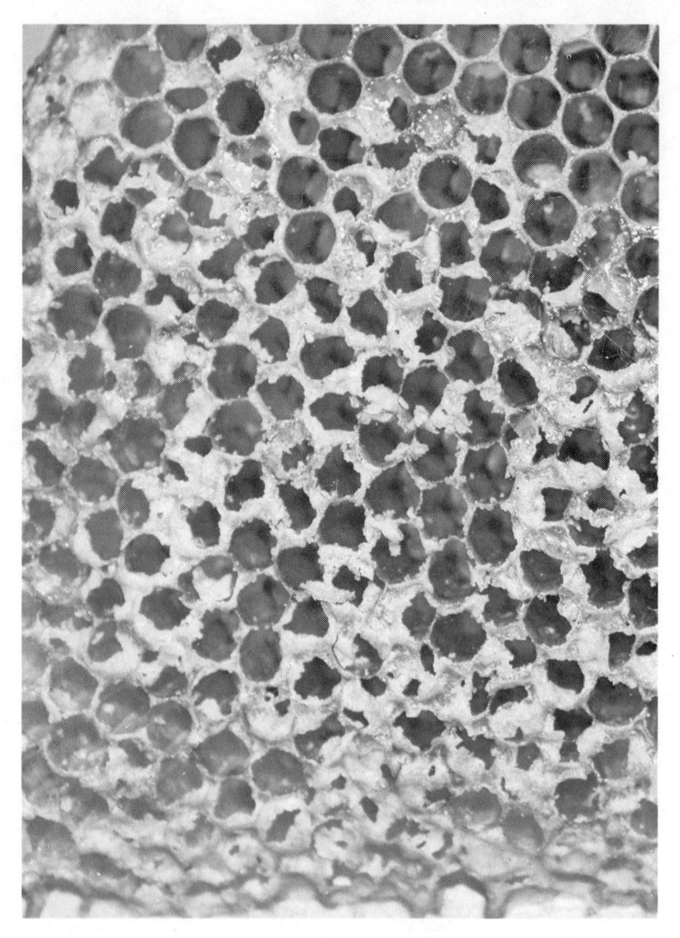

FIGURE 3.4. These shaggy edges can indicate robbing. Credit: Christy Hemenway.

Unlike the orientation flights of young bees, robbing goes on until dark, or until the food supply is gone. Robber bees, once they get started, are in it for the duration. This is why it is such a disaster for the robbed hive—it is very difficult to stop.

If you close up the hive when it gets dark and then watch for activity in the morning, any bees that show up early looking to get into the hive are probably robbers returning from the day before.

Also look for scurrying behavior. Bees that are robbing are in a hurry to get to the food source. They get in at the entrance and quickly go directly to the honey stores. Experience and observation will help you learn the differences between busy bees and running, robbing bees.

Chewed comb is another indication. Robbing bees tear the cappings of the honeycomb cells in their haste. The edges of the cells will be rough and jagged. Pieces of chewed wax cappings will accumulate on the bottom of the hive. Examining the trash found on the bottom board of the hive can help you to diagnose a robbing problem. Be careful not to confuse these shaggy looking cappings with the damage caused by mice; they look somewhat similar.

Causes of Robbing

Oddly enough, it is usually the larger, stronger hives, those with the most stores, that exploit weaker colonies nearby. Italian honeybees seem to have a stronger inclination to rob than other breeds of bees. Robbing behavior generally starts late in the season, when there is a dearth, a lack of nectar in the bees' forage. It is especially likely if the nectar source dried up suddenly. This often happens when the bloom in a large field of a cultivated crop ends naturally, or when it is abruptly harvested, as opposed to the gradual seasonal changeover of the wildflowers and weeds that are the mainstay of bee forage in areas with little commercial agriculture.

The following activities can also encourage robbing:

- Feeding during a dearth. This is a bit of a Catch-22, since not feeding isn't a good option—yet feeding and having it start a robbing frenzy in a hive that needs feeding can lead to disaster.
- Feeding an open source of honey or sugar syrup anytime, especially near your hives' entrances. This invites other bees to investigate.
- Opening hives to inspect them can incite robbing. Bees will be attracted to the open honey stores. This is especially problematic during a dearth.
- Leaving a top bar full of honey outside the hive. This is extremely risky, especially during a dearth. Such easy pickings quickly get the attention of robbing bees.

In the long run, bees may begin robbing just because they can. As the beekeeper, your best course is to minimize the likelihood of it happening at all.

Preventing Robbing

The best possible solution to robbing is to prevent it from getting started in the first place. Here are some ways to do that:

- When the weather goes hot and dry late in the season, make the entrance(s) to all of your hives smaller. This means corking round entrances or installing an entrance reducer in a slot entrance. This minimizes the space that the bees must defend, making it easier for them to turn away any marauders that are nosing around looking for an opportunity.
- Feed your bees inside their hive, not out in the open. Check that syrup feeders are level to prevent leaking, since dripping sugar syrup can quickly attract robbing bees. It is helpful to refill empty feeders in the evening, as the robbers will have returned to their own hive(s) at dusk. Be very careful not to spill any syrup while refilling or replacing the feeders.
- Avoid the use of "Boardman-style" entrance feeders. Though these are useful in end entrance hives that have landing boards, the smell of the syrup can attract robbing bees right to the door!
- When inspecting, be sure to keep your hives buttoned up to the greatest extent possible, especially in the case of a lengthy hive inspection, such as often occurs during classes in teaching apiaries.
- Be sure to keep any honey that you plan to harvest inside a container with a tight-fitting lid while you are still out in the bee yard.

Stopping a Robbery in Progress

Robbing bees will continue to rob until the source of food is completely gone, unless you take definitive action to stop them. A strong hive can rob out a weak hive's stores in a single day. You do not have the luxury of waiting to see if they will give up—they will not stop. So don't lollygag. Act *immediately*!

First: Reduce access to the hive being robbed

- Slot entrances: close slot entrances down to the width of one bee.
- Round entrances: cork all but one round entrance. To make that remaining round entrance even smaller, cut a cork in half and insert one half into the entrance.
- Look for other unintentional entrances that the robber bees may have discovered. Seal these up completely.

Second: Minimize the smell of food

- If your hives have screened bottoms, with bottom boards that can be lowered or removed, be sure the bottom board is installed and in the closed position. This helps to minimize the smell of the hive's honey stores wafting out to entice the robbers.

Third: Confuse the robbers

- Cover the hive with a wet sheet. I know this sounds silly—it looks a little silly too—but it works to prevent the robbing bees from getting at the entrance of the weaker hive. The robbing bees are not interested in solving puzzles or mazes, and soon give up. The bees that reside in the hive don't seem to have any trouble finding their way in, or out. Yes, you will want to wet the sheet again when it dries out, as it is the weight of the water that helps to keep the sheet draped over the hive, concealing the entrance.
- Lean a piece of glass against the hive in front of the entrance. This will also confuse the robbers. The resident bees seem able to find their way around the glass; the robbers, not so much.
- Install a robbing screen, a device designed to block direct, straight-on

FIGURE 3.5. A robbing screen will help to confuse the robbing bees and protect your hive. Credit: Bill Domonell.

access to the entrance. This stops the robbers, yet the bees that live in the hive quickly learn to walk around the screen's edge to enter and exit their hive. The robbing bees are intent upon flying straight at the entrance to enter the hive they are robbing, and they just can't seem to figure out how to go around.

- Another method, according to Michael Bush, is to smear some Vicks VapoRub around the entrance. The robber bees are guided by smell, and the intense menthol odor of this product confuses them.
- The tendency of bees to continue to rob until they have completely consumed the source of food can also be used to confuse the robbing bees. If robbing has already begun, the food source that incited the robbing—in this case your hive(s)—should be moved. However, the robbers will then likely turn their attentions to another hive in your apiary, typically the weakest one. To prevent this, after you have moved the original food source, replace it with a smaller food source, such as some honey or sugar syrup on a plate. A few spoonfuls are usually adequate, and after consuming it, the robbing bees will settle down and depart, believing they have finished.

Other Issues with Robbing

Overheating

Closing your hive up completely in hot dry weather in order to prevent robbing can cause issues with heat buildup and ventilation. Here are some ways to work around that:

- Place a piece of screen over the entrance to lock all your bees into the hive that's being robbed. Install this screen at night, once the robbers have gone home, and your bees are all inside. The screen will help with ventilation, and if you can feed and water them inside for 2 or 3 days, the robbers, who have been locked out by the screen, will generally get tired of trying to get in and give up.
- The opposite technique, which also works, is to close up the hive, again by installing a screen over the entrance during daylight hours, while the robbers are still inside. If you keep them trapped inside, and

provide the hive with food, water and adequate ventilation for about 72 hours, this will eventually force the robbers inside to join the hive they were robbing.

Starvation

Sometimes a weak hive will get completely robbed out, and there will be nothing left of honey stores. This is terribly bad news if it occurs late in the season, as the robbed hive is very likely to starve to death. If you have such a weakened hive, and you also have another, stronger hive, you can combine them. One strong hive always stands a much better chance of surviving the winter than two weak hives. It's best if you can wait to combine them until the conditions that caused the robbing have abated, if possible.

If the colony has been completely destroyed by the robbing, you will want to preserve the comb and prevent mice from moving in and destroying your bees' handiwork. Close the entrances completely but do not seal up the screened bottom. Make sure the screened bottom does not permit entrance to mice. Air and light work best to preserve honeycomb, and winter's cold will destroy many pathogens.

Robbing by Not-Honeybees

Yellow jackets, hornets and wasps have also been known to rob beehives. In general, these insects are stronger and tougher than honeybees, and they can fly in colder temperatures. They are sometimes interested in the bees' honey stores, but more often they are more interested in killing and eating bees and bee larvae. Here are some ways to deal with these creatures:

- Install the type of robbing screen on your beehives as described above.
- Place attractant traps designed for yellow jackets around the apiary. Hang them some distance away from your hives, causing the wasps or yellow jackets to go to the traps instead of pestering your honeybees.
- Move the hive to a new location.

4 Combining Hives for Overwinter Success

Why Combine Hives?

In the fall, beekeepers evaluate the survival prospects of each hive, based on the size of the colony and the amount of stored honey. A colony that has stored less than 6 to 8 full bars of honey during its first season is likely to be in trouble when it comes to overwintering. You could attempt to support a small hive through the winter by feeding heavily in the absence of honey stores, but the truth is that it is easier and less stressful for the bees to overwinter as one large thriving hive than it is for the beekeeper to try to nurse two struggling hives or dinks.

This is a fairly large and traumatic reorganization of the hive, so it's best to make the decision to combine hives early rather than late. If it is done too late in the fall, the bees will be unable to reseal gaps and cracks with propolis, leaving them with a drafty hive, difficult to overwinter in.

Interchangeable Equipment

This manipulation highlights yet another good reason for maintaining interchangeable equipment in your bee yard. The top bars especially must fit in both hives if your goal is to combine the two.

FIGURE 4.1. Warm propolis stretches between the bars when inspecting. Credit: Ben Sweetser.

FIGURE 4.2. A bee returning to the hive with the tree resin that will be used to make propolis. Credit: Ben Sweetser.

What About the Queens?

If you have a strong preference for one of the queens, you might first choose to purposely make the other hive queen-less, then move the bees into the same hive. In a situation where the two hives are disparate in strength, the queen of the stronger hive would be the queen to choose to keep; otherwise the choice is up to you. One could simply combine the hives, leaving the choice of queen up to the bees. But this runs the risk of having both queens be killed or injured, usually occurring at a time of the season when replacing a queen is an arduous task for the colony.

Organizing the Interior of the Hive When Combining

The configuration of the honey stores inside a top bar hive is important for overwintering success. As was shown in the chapters of *The Thinking Beekeeper: A Guide to Natural Beekeeping in Top Bar Hives,* which detail the specifics of top bar hive management, the hive's comb building and expansion needs to have been shepherded so as to be *unidirectional.* In other words, the beekeeper should manage the initial building of comb so that it progresses from brood to honey stores, preventing the bees from putting honey stores on both sides of the brood nest. It is crucial to understand the reason for this: While the temperature dictates that the bees must stay in their winter cluster, individual bees cannot move away

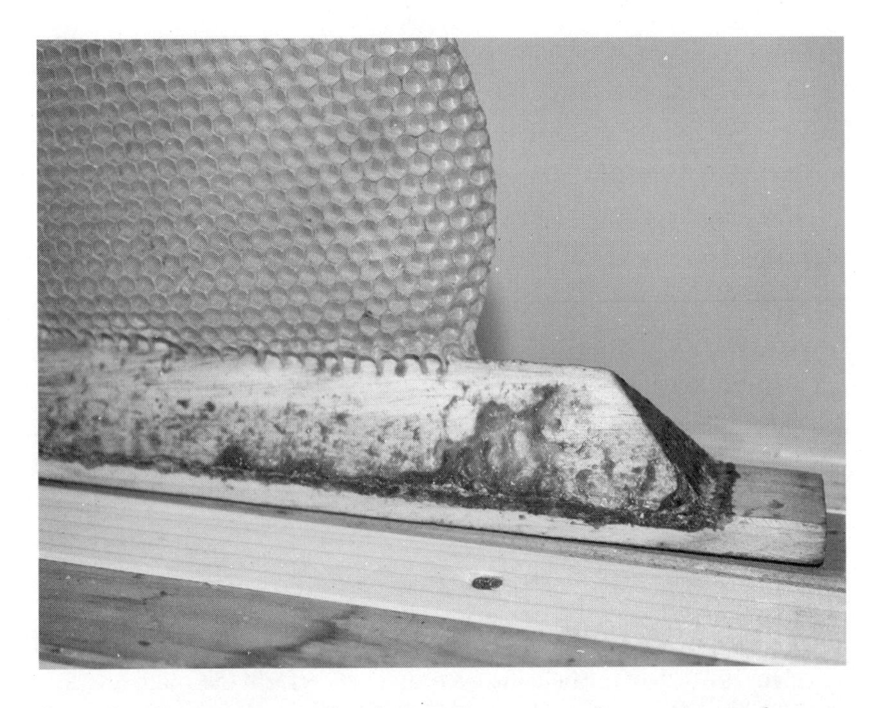

FIGURE 4.3. The bees close up gaps and glue everything together with propolis. Credit: Christy Hemenway.

from the cluster, nor can the cluster move across dry empty comb to get from honey stores on one side of the hive to honey stores on the other. To help your bees survive, manage them so that their honey stores are all on one end of the hive.

This management process is different depending upon the type of top bar hive used. End entrance hives begin at the front wall of the hive; side end entrance hives begin at the wall on one side. In these hive styles, the building of comb progresses from a fixed wall of the hive and moves toward the opposite end. In a side center entrance hive, having two movable follower boards, this progression is guided by conscious management on the part of the beekeeper, and begins at a movable wall (one of the follower boards). When installing a package in a side center entrance hive, the beekeeper will *anchor* the colony at one end of the *bee bowl* by placing the queen cage on the bar directly against one follower board, thus allowing the bees to build comb in only one direction, away from the anchor bar. As the colony grows and needs to be shifted mid-season, the entire hive contents are then shifted down to one end of the hive so that all the empty top bars can be relocated to the other end—the end that the hive is building toward. Voilà! Unidirectional bees! This management prevents the bees from storing honey on both sides of the brood nest and setting up a potentially devastating situation in the winter.

These factors all come into play when one plans to combine hives to increase their chances of surviving the coming winter. Integrating the contents of two hives presents some particular challenges when the bees are able to make their own wax as they do in top bar hives, since the shape of each individual comb is built to match the comb next to it. Shuffling bars of comb between hives disturbs the integrity of this structure. To minimize the disturbance, the beekeeper wants to remain sensitive to the order of the combs, all the while rearranging things in order to integrate the two colonies successfully.

How to Do It

To retain as much of the field force as possible, perform this manipulation late in the day, so that as many bees as possible have returned from the field and are in the hive, on the combs that you are moving.

To begin, create a paper divider to separate the bees inside the hive when they are first moved. Use newspaper, or other lightweight paper, that the bees can easily chew through in order to join up. The follower board from your hive is the perfect template to use for this divider, since it needs to match the shape of the interior of the hive body.

FIGURE 4.4. Newspaper divider technique. Credit: Geoff Keller.

Install the paper divider at the new comb end of the hive you are moving the bees into. Do not include any blank bars or tiny combs from the original hive; use only fully *drawn combs* or combs containing brood or food. Tape the divider to the sides and bottom of the hive body interior; the top can be folded over the top of the last top bar and held in place between top bars. Cut two or three one-inch slits in the paper to facilitate the movement of pheromones between colonies and to provide a starting place for the bees to make their way through. They will chew their way through the paper over the course of the next few days.

On the far side of the paper divider, add the bars of comb from the hive you are moving bees from. Move brood combs first, then honey combs. Place small combs at the new end. Include any small combs from the original hive here. Fill in the hive with the rest of the top bars. Always be sure that all of the top bars are in place in your hive when you are finished.

Getting the Foragers to Move

Your next challenge is to move the field force—the foraging bees. Many young bees will have moved with the combs; what you need now are the bees that were out foraging when you made this move and combined the two hives. If possible, place the hive you have moved the bees out of close to the hive you have moved the bees into, with their entrances facing each other. The bees stranded in the old hive will discover their sisters have left and, following the smells and pheromones, will move to the new hive on their own.

Integrating the Shuffled Comb

Close up the hive and check back in several days. By this time, the bees should have eaten their way through the paper and joined themselves (more or less) into a single colony. Note that there may still be bees on young brood on the combs you moved, even though there may be empty or only lightly covered bars between the brood combs of the existing hive and the brood combs of the relocated hive.

Now, you can shift the individual combs to integrate the two hives in an order that moves from brood to honey, while striving to recreate the unidirectional progression inside the hive. Using your best judgment, try to place combs of similar purpose and size next to each other. In the case of combs that will not nest snugly against one another, try a different location within the order of the combs, or try turning the bar 180 degrees. If you have combs with wide "shoulders" at the top, insert a spacer between the combs to keep them from touching or crushing each other.

Leave the bees alone to sort themselves out before winter arrives. As long as they have integrated well into the new hive configuration, and they have adequate food stores, there is little reason for further inspections at this time in the season. You have done what you can. Check their food supplies, if you can, on a sunny winter "thaw" day and feed fondant then, if needed.

5 Honey—Sweetness and Light = Food!

HONEY IS A MAGICAL ELIXIR. Sweeter than most anything on Earth, versatile in cooking, potent for healing, delightful in tea and the only food that is known for never spoiling or going bad...this amazing product of the hive is one of nature's marvels.

Bees produce honey to be consumed as food, especially during winter. It is absolutely crucial to their survival during the time of year when there are no flowers to forage among to collect nectar and pollen.

Honey is the product of a long-standing relationship between flowers and bees. Flowers need to be pollinated to reproduce; bees need food. The flowers have devised all sorts of interesting ways to attract bees and other pollinators in order to move their pollen around, and the bees use the nectar they gather from the flowers to make the ambrosia we call honey.

All of this activity is what causes the plants to make food for us as well...and the number of important connections this creates between bees and humans and our food system and agricultural methods is mind-boggling!

It is the bees' ability to make more honey than they require to live on that beekeepers exploit when they harvest honey. This makes it incumbent upon us as beekeepers to be circumspect about how we manage this finite and precious resource. Taking honey from the bees should be done thoughtfully and conservatively. There is little long-term advantage to anyone, especially the bees, if the beekeeper removes all of the bees' natural food and replaces it with a nutritionally poor substitute such as sugar syrup, or worse, high-fructose corn syrup.

Taking this long view, we often say, "It's not about the honey, Honey—it's about the bees." Healthy bees make honey without being forced to, provided the beekeeper supports their natural systems and does not manipulate the bees in detrimental ways.

FIGURE 5.1. Bees capping cells filled with ripe honey. Credit: Ben Sweetser.

How Much Honey Do the Bees Need to Make It through the Winter?

I spent a long while observing top bar hives and top bar hive beekeepers before I decided there was a reliable answer to this question. And it still depends, of course, upon many factors that are specific to your location: the most obvious being the size of the colony, and the length and severity of the local winter.

The bees' ability to store honey is also impacted by the length of the foraging season or *bloom*, the weather during that season, the amount of *forage* in their locale and the health and vigor of the colony. After combining my own experience in Maine with that of other successful top bar hive beekeepers throughout the US and

FIGURE 5.2. Capped honeycomb is ready for harvest! Credit: Christy Hemenway.

abroad, it seems fair to say that 6 to 8 full bars of capped honey will sustain a top bar hive through a "typical" winter. A colony that has that much honey likely also has several bars of brood comb that have been backfilled with honey as well, and together, this seems to do it.

Typically a top bar hive beekeeper waits until the spring of the hive's second year to harvest honey, leaving the bees as much of their own natural honey as possible to overwinter on for their first winter. This is a good practice to follow, since honey is, after all, what honeybees need to eat to get them through the winter! In their first year, a new colony has plenty to contend with just to get started as a colony and prepare to survive that first winter.

But in the spring of their second year, if there is still honey in the hive, and there are blossoms in the field, one might be so bold as to consider some of the bees' surplus honey to be your own reward. And such a superior honey it is, too. Honey produced by healthy bees doing their own thing on their own clean natural wax is some of the cleanest and healthiest and best honey you could ask for. It's also as local as it can get, being from bees that live in your own backyard. And if you don't heat it when you harvest it, then it remains as full of rich natural goodness as it can possibly be.

Yes!!! A Harvest!

So let's say your hive has come through the winter with flying colors, and even survived April, that cruelest month in beekeeping, and now you are back into bee season. There is plenty of forage, and there is capped honey still in the hive as well. It's spring, probably May, possibly June; it will depend on where you live. This is the time of the year when it makes the most sense to remove honey from your hive.

The season still stretches out before you. Harvesting a moderate amount of honey at this time allows the bees to replace their stores during the summer foraging season. Harvesting most of the hive's honey late in the season deprives the bees of their stores at a time when they cannot reliably replace them, and too often this is the reason beekeepers must feed to prevent starvation.

Getting the Honey out of the Hive

So how do you get the honey out of the hive? Gather up an appropriate container that is large enough to hold a fully drawn comb and has a tight-fitting lid. If you prefer to keep your harvest as cut comb honey, be sure that your container is large enough for the combs to lie flat without bending or breaking. You will also need a good sharp knife. If your top bar hive tool is a long, straight-bladed knife, then that's a good tool to use for this purpose. Wear your protective gear. Remember, you are removing their life's work and their winter food supply—perhaps it's understandable if they are a bit defensive about your presence in their hive.

Choose the combs for your harvest judiciously. Choose the light-colored combs that the bees made for honey storage, not their brown brood comb. Remove only bars with all (or as close to all as possible!) the honey capped. Nectar that is not yet capped has a moisture content that is higher than honey, and this can cause the honey to ferment after

FIGURE 5.3. These are baby bees—not honey. Don't harvest them! Credit: Christy Hemenway.

harvesting. Not that uncapped nectar or even slightly fermented honey is necessarily a bad thing for human consumption, but it tends to expand in the jar after bottling, which can lead to leaks and a sticky mess, or even broken jars if it is stored for any length of time.

Remove the bar you've chosen to harvest from the hive. Hold the comb over your hive, and use a bee brush or a handful of grass to brush the bees from the comb.

Saving Honeycomb on the Bar for Winter Bee Food

You might want to remove some bars of honey in anticipation of it being surplus, but wait to remove it from the top bar in order to see whether the bees might need it later in the season, during a summer nectar dearth or in preparation for winter. If you leave the comb attached to the top bar, it is a simple matter to place it back in the hive if the bees should need it.

Since the bars in a top bar hive must all touch in order to maintain the proper spacing, and they must all be in place at all times to prevent your

FIGURE 5.4. **Remove the honeycomb from the bar. Return the bar to the hive!** Credit: Christy Hemenway.

FIGURE 5.5. **Gorgeous cut comb honey.** Credit: A & R Verbeek.

bees from building comb down from the inside of the roof, you should plan to have some spare top bars on hand.

Harvesting Cut Comb Honey

Use a sharp knife (AKA your hive tool) to cut the comb away from the bar. Place it gently in the container. Be sure to always cut away from yourself as you remove the comb from the bar, as the knife may jump a bit as it cuts through the comb, and it can seriously damage beekeeping gloves and fingers.

Replace the top bar in the hive while offering a word of thanks to the bees. Repeat this process for each bar that is appropriate to harvest at this time. Be sure to put the lid on your container between choosing bars, as the bees will certainly be interested in going in after their food, AKA your harvest!

To store cut comb honey, remove the comb from the hive, cut it carefully from the bar and simply cut it into pieces that will fit the container(s) you have in mind for storing it. Choose something unique and attractive for an impressive presentation. You can add some liquid honey to the container as well, if you like.

Oh, the taste of honeycomb fresh from the hive! An amazing sweet summer treat, comb honey can be eaten just as it is. A chunk of comb honey melting across a piece of warm toast is one of the most amazing things to come from a beehive.

Liquid Honey

Harvesting *liquid honey* from a top bar hive is typically done via a simple method called "crush and strain." This low-tech method of collecting honey in its liquid form means doing just what it says—crush the wax combs and strain the honey.

A Gold Star Honey Harvest Kit is made from two 5-gallon buckets and their lids. This is large enough to accommodate approximately six combs at a time. The bottom bucket acts as the collector bucket. Drill a 4-inch hole in the center of its lid. Install a *honey gate* in the side of this bucket as close to the bottom as possible; it will make bottling the honey much easier.

The top bucket will hold the strainer and the honeycombs that you are preparing to crush. Drill a series of approximately 18 half-inch holes in concentric circles within a 4-inch circle in the center of the bottom.

When the top bucket is placed on the lid of the collector bucket, the holes in the bottom of the top bucket will align with the large hole in the lid of the collector bucket. The honey will drain through the strainer and into the bottom bucket.

Now you need a strainer. In the case of a 5-gallon bucket, a commercial paint strainer bag will work nicely. They are available in the paint department at the hardware store. The strainer will line the bucket, and is fine enough to strain out beeswax and bee parts, while allowing the honey and its beneficial pollen to pass through.

FIGURE 5.6. **A Gold Star Honey Harvest Kit makes it easy to harvest liquid honey.**
Credit: Brian Fitzgerald.

Important Note: Do not heat honey during the harvesting process. Heat destroys the beneficial enzymes and nutritional elements of the honey, and affects its desirable qualities, such as smell, taste and color.

FIGURE 5.7. **All the cells containing honey must be broken open!** Credit: Christy Hemenway.

Next you need to crush the combs. Add the harvested combs to the strainer in the top bucket, one at a time. Crush the comb with a potato masher, a wooden spoon or simply squeeze pieces in your hand. Avoid using anything sharp that can cut or puncture the strainer bag. After completely crushing all the combs, put the lid on your top bucket and set the honey harvest kit in a comfortably warm place. Let gravity do its work. This can take the better part of several days.

That's all it takes to harvest liquid honey. Be sure to stir and turn the contents of the strainer bag to help the honey drain. As a final effort, you can twist the top of the strainer bag and squeeze it to force out as much of the remaining honey as possible.

Let the honey sit in the bucket for a day or two. You may find that some "foam" will rise to the surface. The foam is perfectly edible but is not usually bottled with liquid honey, especially if the appearance of the honey is of concern.

Now the honey is ready to be bottled. Glass containers are best for storing honey. Glass is clean, benign and does not impart any taste to the honey, as plastic can. And unlike plastic, glass is heat resistant. Should the honey crystallize, a common and harmless occurrence especially in the wintertime, it can be made liquid again by setting the glass jar in a pan of hot water.

Do not use a microwave oven to warm crystallized honey, especially if it is in a plastic bottle. Also do not set the jar in a pan of water still boiling on the stove—that will pasteurize the honey, as discussed earlier.

Now that the honey has been harvested, you will find yourself with a strainer containing a mass of crushed and broken pieces of beeswax. Rinse the wax in the strainer under room temperature running water to remove as much of the remaining honey as you can. At this point the wax is ready for rendering (discussed in the next chapter). Store it in your wax box until you're ready to render it.

Benefits of Crush and Strain Harvesting

Harvesting honey by the crush and strain method is simple and low-tech. The honey is not heated, nor is it forced through an ultra-fine strainer. This method retains the important healthful benefits of pure natural honey, including pollen, which is what gives honey its own special *terroir*—that particular set of characteristics such as taste, color and smell—that is specific to the locale where it was made.

The natural clean beeswax harvested by the crush and strain method has multiple uses. Clean beeswax from an untreated hive is superior for making personal care or cosmetic items, such as lip balm or hand creams; any oil-based herbal preparation is thickened by adding beeswax. Beeswax is also used for making beeswax candles, far superior to paraffin candles.

Honey as Allergy Medicine

Those who suffer from a pollen allergy may find relief from their symptoms by eating local honey. This is why locale is so important when it comes to honey; only honey produced where the allergy sufferer lives will have been made from the flowers that cause the allergy. This means your backyard honey is a powerful natural antidote for pollen allergies.

It is important not to strain the honey through a filter so fine that this very important ingredient, the pollen, is removed. The filter does not need to be any finer than what is required to remove stray bee parts and crushed wax, typically 600 microns.

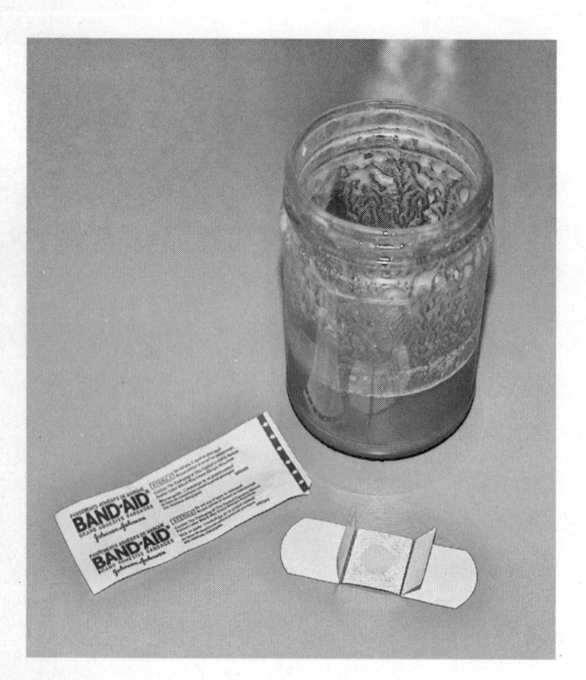

FIGURE 5.8. **Honey can help heal wounds and burns.** Credit: Christy Hemenway.

Honey for Healing Wounds

Honey applied to a cut, scrape or burn can help it heal more quickly and lessen scarring. Of course it's also very sticky, so you'll need to cover it with a bandage.

Why Does the Honey Look Different throughout the Year?

The appearance of capped honey can vary throughout the season. Variations in color are caused by the different nectar sources the bees were foraging on. It can be as light and clear as water or as dark as a good stout. Early in the season, honey is typically light and clear, somewhat thin and tends to crystallize quickly. Later in the season, it is often darker, thicker and stronger in taste. You could literally follow the progression of what is in bloom in your area by tracking the changes in color and taste of the honey your bees are making from local nectar gathered from what is currently blooming.

Dry Cappings Versus Wet Cappings

When the bees have evaporated the moisture from the nectar stored in the honeycomb cells to the level where it is considered *ripe honey* (approximately 18%), they then *cap* the cells. This is like putting a tiny wax lid on each cell. The cappings are described as dry or wet. A *dry capping* has a tiny airspace behind the cap, so the honey does not touch the back of the cap, and it looks like a dry white wax lid, with no sign of the liquid inside. In a *wet capping*, the ripe honey touches the back of the wax cap, causing the color of the liquid honey to show, and giving a translucent, glossy or wet appearance.

FIGURE 5.9. Dry cappings appear white and opaque. Wet cappings are more translucent, and you can see the color of the honey. Credit: Christy Hemenway.

Feeding Honey to Your Bees

It would seem like a no-brainer to feed honey to bees, especially as opposed to feeding them sugar syrup. However, there is a potential risk involved. If you are a treatment-free beekeeper, and you are able to feed your bees honey that you are certain comes from a hive never treated with antibiotics to prevent American Foulbrood (AFB)—perhaps from your own treatment-free hive—this is a fine idea.

However, honey harvested from hives that have been treated for AFB may contain the spores that cause AFB, even though the hive was unaffected and showed no symptoms. This could cause your bees to contract AFB, one of the deadliest and most contagious of bee diseases. Be very, very confident about the source of any honey you choose to feed to your bees.

6 About That Wax That It's All About

Beeswax is the light-colored waxy substance that honeybees use to form their comb, which makes up the internal structure of the hive. Comb is what houses the brood and the food stores of nectar and pollen.

All worker honeybees make beeswax. Worker bees between 10 and 20 days old are best able to produce wax. After this age, their wax glands decline and they move on to other tasks, such as foraging. Transparent liquid wax is secreted from eight wax glands found on the underside of the bee's abdomen, which then harden into eight small oval wax scales upon contact with the air. A bee can produce eight scales of wax in twelve hours. Bees use their legs to move the wax scales forward to their jaws, where they *masticate* the wax, adding secretions from their saliva to create the right consistency.

The ambient temperature in the area where the wax is being built must be at least 91°F (33°C). To obtain this temperature, the first combs are built inside the cluster of bees, where they can create these temperatures. The wax-building bees *festoon*, or *bridge*, hanging in chains while they are secreting wax. When a bee's wax glands have secreted eight scales, she climbs up the chain to the place where the comb is actively being constructed, and contributes her eight wax scales to the building process.

FIGURE 6.1. Bridging or festooning is how the bees work with gravity to build natural beeswax comb. Credit: Ben Sweetser.

FIGURE 6.2. Bridging bees can hold on incredibly tight! Credit: Ben Sweetser.

The construction of beeswax comb is an impressive feat. The hexagonal shape of the cells uses the smallest amount of material possible to create the highest strength and the most efficient use of space. The cells that are attached to the top bars of your hive, which hold the comb and all its contents, can hold more than 1,300 times their own weight. Their six equal sides leave no wasted space between cells. Pretty good engineers, these bees!

Beeswax can vary in color from pale yellow through orange and red to brown and grey. The color makes no difference in the quality, although lighter-colored wax is usually considered more desirable. When comb is used to raise brood, it turns quite dark over time. This color is due to the cocoons spun by the bees as they pupate, which are left behind when the young bees emerge from their cells.

The chemicals used by some beekeepers to control the varroa mite adversely affect the quality of beeswax. The commercial beekeeping industry has succeeded in contaminating 98% or more of the beeswax supply that is used to make foundation. This means that, for your bees, the clean natural combs in your untreated, natural wax top bar hive is pretty special stuff.

How to Save and Store Beeswax

There comes a time for most beekeepers when they find themselves with a loose, broken or even a purposely detached piece of natural beeswax comb in their hand, wondering what the heck they should do with it. It's obviously pretty important stuff, but since "this little bit in my hand doesn't add up to much," it sometimes gets discarded…sometimes in the trash, sometimes on the ground. What a waste! Especially if it is clean, natural wax from a hive that has never been treated with chemicals!

In the interests of keeping a clean and healthy bee yard—one that looks good, feels good, thrives, and doesn't attract pests—and especially in the interests of not throwing away such beautiful and valuable stuff, I suggest that everyone should have a *wax box*. What makes a good wax box? A sturdy container with a tight-fitting lid—that will hold the bits and pieces of broken comb that you will find yourself collecting until you are ready to *render* them.

You may be surprised how quickly beeswax accumulates. How long you wait before rendering the wax may depend upon what you'd like to do with it. If your wax collection grows slowly, and your wax box is small enough to fit in your freezer, that's a good place for it to live—at least for a couple of days at a time—because freezing the wax will eliminate any wax moth larvae that may be hiding there. If you need a bigger wax box than will fit in your freezer, be sure that the lid fits tightly and don't let it sit around for months; render it promptly to prevent wax moths getting into it!

FIGURE 6.3. A good choice for a wax bin. Credit: Christy Hemenway.

The bees do not gather wax in the same way that they gather nectar and pollen and propolis, so you won't find them picking up bits of wax and taking them back into the hive to create comb. All the broken bits and pieces of comb from your hives should go in your wax box.

Rendering Your Beeswax

If you are a treatment-free beekeeper, your wax is some of the cleanest, purest beeswax available. All of the wax from your hive—brood comb and honey comb—can be rendered successfully; they will both yield wax. Once rendered, the wax turns into a solid block of beeswax, which can be stored with no threat from wax moths. Here are some suggestions for rendering your clean, natural beeswax.

Solar Wax Melter

The beauty of a solar wax melter is that once you've built it, the hard work is done. Happily, a solar melter requires no fuel in order to render wax; the sun provides all the heat needed for free! The temperature inside the wax melter only needs to reach 155°F (68°C).

At its simplest, a solar wax melter is a box that collects heat, so it is often painted black inside. To let in the sun, it needs a clear glass or plastic lid, one that seals tightly to contain and amplify the heat. The box needs to be big enough to hold two containers: one that you will place your saved combs into, and another one set below it. This bottom container should contain water, which will catch and cool the clean beeswax after it has melted, and keep it from sticking to the container. Between the two containers, the wax should flow through a strainer to remove debris.

FIGURE 6.4. Comb in a solar wax melter—ready for the sun!
Credit: Christy Hemenway.

How to Do It

Place the wax comb in the top container. Be sure the strainer is in place. Add water to the bottom collector container. Put the clear cover in place.

Reposition the box throughout the day so that it is always facing the sun, and maintain at least a 40° angle to the direction of the sun. The wax you render in this way will have soft rounded edges and an amorphous blobby shape, due to the fact that it dribbles from one container to the other.

There are plenty of plans of varying complexity available on the internet.

Boiling Method

Another way to render your wax is to boil it. This requires more labor for the beekeeper, but it also takes less time, and of course it works even on cold and cloudy days. For this method you will need:

- a *wax pot*
- a strainer
- a heat source
- a 5-gallon bucket or similar heatproof container
- a stir stick

Your wax pot will never again be suitable for anything other than rendering wax, so be sure the pot you choose is not a family heirloom. Look for stainless steel or enamel, as beeswax can react with other metals such as aluminum. Secondhand stores often have good wax pots.

While you're there, look for a large strainer as well. A coarse metal kitchen strainer is a good choice. The closer the size of this strainer is to the diameter of a 5-gallon bucket, the better, as this will help prevent it from falling into the bucket while pouring. A stout piece of wood such as a 1" by 2", longer than your wax pot is deep, can be used for stirring the pot as the wax is melting.

You will need a substantial source of heat, as you will be boiling a significant quantity of water and wax in your wax pot. Since this can be messy work, it is a good idea to do this job outside using a heavy-duty camp stove or similar appliance, but if you're careful, it can also be done on your kitchen stove. If you do process your wax in your kitchen, be sure to protect the floor with a layer of plastic such as an old shower curtain; otherwise, you may find yourself scraping small droplets of beeswax off the floor long after this project is complete.

FIGURE 6.5. A good choice for a wax pot. Credit: Christy Hemenway.

FIGURE 6.6. This wax pot will never be good for anything but a wax pot! Credit: Christy Hemenway.

How to Do It

Fill your wax pot about ⅓ of the way with water. Bring to a boil. Add the combs one at a time. If you have both brood comb and honey comb, you might choose to render them separately to preserve the light color of the honey comb.

Do not fill the pot with comb more than ⅔ full, to prevent hot wax and water from splashing over as the water boils. Stir to get all the combs down into the water. The wax will melt and disappear into the water.

When all the combs have melted, set your strainer into a 5-gallon bucket. Protecting your hands, carefully lift the pot full of hot water and wax and pour it through the strainer into the bucket. Leave it to cool for several hours. The more slowly it cools, the fewer impurities become trapped inside the wax.

The wax will float to the top, hardening as it cools. When it is completely cool, run a sharp knife around the inside of the bucket to release the wax cake from where it may have stuck to the side of the bucket. Push down on one edge of the round of wax, and the other edge will rise up out of the water. Grasp this edge and remove the wax from the bucket.

There will likely be a layer of fine debris on the underside of the wax cake. How much debris depends on the initial cleanliness of the wax and the gauge of the strainer. Most of this layer can be scraped off; any that remains can be removed by a second rendering in boiling water. For the second rendering, line your strainer with several layers of cheesecloth laid at 90-degree angles.

After the second rendering, the wax cake can be carefully melted and poured into containers that will give the desired size and shape to your wax cake. Coat the container with a very fine film of soapy water. The soap will help to release the wax from the container after it has cooled. Let the wax sit for a full day, so that it is completely cool before attempting to remove it.

FIGURE 6.7. **A layer of fine debris remains on the underside of the cooling wax cake.** Credit: Christy Hemenway.

FIGURE 6.8. **A cake of rendered beeswax.** Credit: A & R Verbeek.

FIGURE 6.9. Beautifully rendered beeswax. Credit: A & R Verbeek.

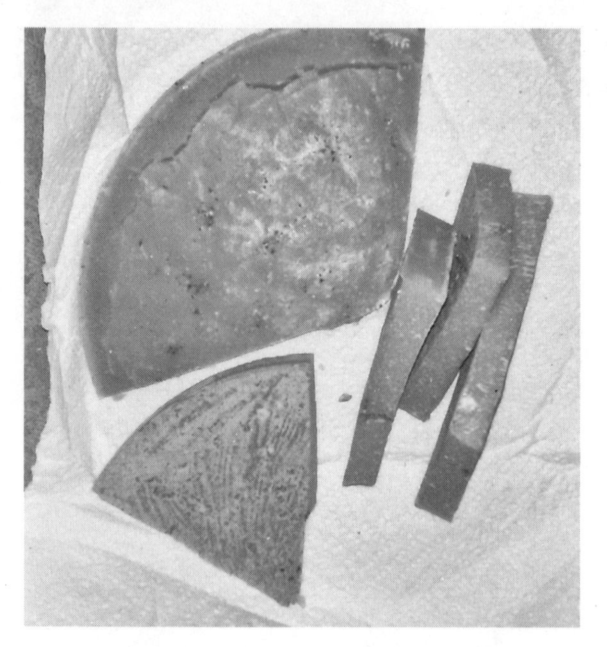

FIGURE 6.10. Clean beeswax ready for sale or for your projects! Credit: A & R Verbeek.

Fire Starters, or... Waste Not, Want Not

The dark brown cocoons left behind when brood comb is rendered will collect in the strainer. Don't throw these away! They still contain a significant amount of beeswax and can be used to make fire starters for your fireplace or woodstove. Use the cardboard tubes from paper towels or toilet paper, and pack the cocoons and other dreck left behind in the strainer into them while still warm. When they have cooled, use a sharp serrated knife to cut them into 1-inch slices; use them to help start your kindling on a cold winter night.

Safety note: Notice that both methods of rendering use water. When rendering wax by boiling it, this increases the safety of the process of melting beeswax, which is flammable and can be very dangerous to heat directly over a heat source. Water and beeswax have a unique relationship, the most important feature of which is that beeswax floats. The liquid wax will be indistinguishable from water while it is hot, but as it cools, the wax floats to the surface and becomes the yellow color that is the hallmark of clean rendered beeswax.

Some Uses for Beeswax

Soaps, Lotions and Cosmetics

Beeswax can be added to any oil-based preparation that needs to be made thicker or firmer. Until you are familiar with your recipe, add small amounts of beeswax at a time and allow the mixture to cool after each addition to test the consistency of your creation. If it is not thick enough, gently reheat and add a bit more beeswax. The addition of any essential oils you plan to use should be done last, as they are adversely affected by repeated heating. The oils will also add a small amount of liquid back into your mixture, so be sure to compensate for that.

Beeswax Versus Paraffin

Beeswax is an organic substance—it comes from the body of the honeybee. Paraffin, on the other hand, is a petroleum product. Beeswax burns brighter and more slowly than paraffin, ionizing the air when it burns, cleansing it. Paraffin emits toxins and plenty of black soot. Beeswax candles burned in a still environment rarely drip or run. Paraffin drips frequently. So while both are called "wax," the differences are significant.

Candles

One of my favorite things in the world is a hand-dipped beeswax taper candle, but hand-dipping candles takes a large amount of wax. Molded tapers and figurine candles require less wax. Herbal preparations, such as salves and lip balm use even less wax, unless you are producing them in large quantities.

Other Handy Uses

A cake of beeswax can be useful in many ways and places. Run the threads of a woodscrew along the edge of the cake of beeswax, and the screw will drive into wood more easily. Rub a cake of beeswax along the runners of a stubborn dresser drawer, and it will slide more smoothly. When sewing by hand, run the length of the thread over the corner of the beeswax cake—this will strengthen the thread and help to keep it from tangling.

7 Year Two: Your Best Resource Is Natural Beeswax Comb

Year Two and You've Still Got Bees...

The sun came out and the temperature got warm, and you went out to look at your hive(s) and found to your delight that your bees have made it through the winter!

Awesome! You've successfully overwintered bees in a top bar hive—something you may have heard could not be done! But it can, and you have done it. This is definitely a cause for celebration. It is also early notice that you should be prepared for a swarm and be ready to perform intentional splits for increasing the size of your apiary, or for helping another top bar beekeeper get started. In any event, you should consider investing in some additional hive equipment. You are likely to need it and probably somewhat sooner than you might have thought.

Now you're looking at this hive filled with *survivor stock*—locally overwintered honeybees. What should you do first? If the growing season is in full swing in your bee yard, and your bees have still got honey, then you might consider a honey harvest for yourself. I would start with an inspection to investigate their honey stores.

If the weather is warm enough, say 65°F or warmer, do a thorough bar-by-bar inspection. Look for all the things you want to see in a thriving hive: eggs and larvae, capped brood, pollen stores, honey, the queen. Take some notes about what you see, and be sure to include the date of your inspection. Now you have a new starting point, and while it's not quite the same as counting from the day you hived your first bees, it is important to know and will still give you a good idea of how the "Bee Math" works so that you can assess their progress during the course of the season.

Check to see what you can most logically do to keep your bees working in a "unidirectional" fashion, just like you did last year. In a side center entrance hive, you want them to go from brood nest on one end to honey stores on the other so that, come next winter, they will once again have their honey stores on only one end of the hive. This prevents the problem of them having to move across empty comb in order to reach additional bars of food that—you probably realize by now—they cannot do when temperatures require them to be in cluster. The cluster must always be in contact with food. Since you were able to inspect them last, the bees may have put honey in places your management methods would prefer that they didn't. See if you can arrange the combs that contain honey so that they are all together on the honey end of the hive; if they won't fit well against the bars at that end, then harvest the honey on one end, leaving the honey on the other.

As soon as practical, monitor your hive for varroa mites. Please don't let them fool you. Just because you don't actually see them, that doesn't mean mites aren't in there. They don't spend a lot of time walking around in the open where you can spot them. Most of the time, they are hidden away in capped brood cells where they reproduce; otherwise they are tucked in between the articulated segments of the bees' abdomens, where it is easiest for them to get the lymph of the adult bees, which they feed on.

If levels warrant it, treat for mites. In *The Thinking Beekeeper*, we talked about how to do a powdered sugar treatment. In Chapter 12, we will discuss the use of *oxalic acid* (OA), which may be something to consider as well.

Culling Comb for Maintenance

So now you've got this hive containing beeswax combs. They're made from clean natural beeswax because your bees made it all themselves—you didn't need to use any foundation, and you never added any toxic chemicals to the hive.

However, it's important to remember that beeswax is *lipophilic*. That means that the beeswax will absorb any fat-soluble chemical elements that it comes into contact with, such as *miticides*, which is how the world's supply of foundation has become contaminated. And it's why we say, "It's all about the wax!"

Now, you know that you're not purposely putting chemicals into your hives. But what if toxic pesticides are getting there some other way? If there is a substantial amount of *Big Ag* within 5 miles of your hives, you can be sure that your bees have been foraging there. And if these crops were sprayed or treated with toxic chemicals, then those chemicals will come back to the hive in the nectar and pollen the bees collected and be absorbed by the natural beeswax combs.

This makes it a good practice to *cull* comb as a part of your long-term maintenance and *integrated pest management* program. Early in the season, before the queen has gotten a good start laying eggs, remove the oldest 2 or 3 bars of brown brood comb from the hive and cut the wax from the top bar. Put the now empty bars back in place at the new end of the hive, in the direction the hive is growing toward. Move the next bars with existing comb up to the starting or anchor bar position.

Culling older comb in this way will help to minimize the accumulation of toxins in the comb that bees may have introduced due to their normal foraging. Save this comb in your wax box for rendering later. In Chapter 6, we discussed ways to turn this culled comb into beautiful beeswax.

Year Two and You've Got No Bees...

Remember how we talked in Chapter 2 about your bees' strong drive to reproduce by swarming and how that was a "Congratulations, I'm sorry" kind of event?

Well, losing a hive during its first winter is a little like that, only different. In this situation, it's an "I'm sorry, congratulations" moment. It's very disheartening to lose bees, whatever the cause. It's especially tough to see them get through the winter but then not manage to make it through the spring. That's why April is said to be the cruelest month in beekeeping. It's so exciting to see that your bees are there and flying in March and then so incredibly sad to discover they've died in April.

In any event, you're standing there looking at a dead-out hive, and wondering what to do next. There are probably some dead bees in the bottom of the hive, and in and on the comb. There might be some mold, pollen, or the white dross that is left behind when uncapped honey/nectar crystallizes in the cells. There may also be white specks at the tops of cells, which are mite feces and indicates the presence of varroa mites. It's depressing.

And I'm sorry. Really, I am. I hear stories from novice and experienced beekeepers alike, frustrated, downhearted, and crestfallen over their bees dying. And believe me, every single one of those stories hurts; it does not get easier, and you do not get used to it, and every time I hear such a story, it's like attending a little funeral.

But even if you have not yet succeeded in overwintering your bees, please don't lose heart. Because... and really, I mean it—Congratulations! You have a precious resource this year that you didn't have last year, and that is natural beeswax comb, made by bees for bees, without the use of foundation and never subjected to the typical chemical treatments placed in hives. This is really very special stuff. You might say it's worth its weight in gold. (Unfortunately it doesn't weigh very much, but still!)

Another thing you've gained, even though it appears that you've been unsuccessful, is a full year of experience with bees. At this point you probably aren't even aware how much you've learned! Take a little credit for the journey you've made thus far.

And... once again—Congratulations! The natural wax you are sadly examining is about to make a very big difference to your beekeeping. Do *not* destroy the comb! Unless wax moths or small hive beetles have in-

FIGURE 7.1. **Having drawn comb will make a big difference to your new bees!** Credit: Christy Hemenway.

fested the comb (both destroy the comb and leave it obviously unusable), or your bees died of a known disease, you now have the stuff to give a new hive of bees a very nice head start. If you can find the fortitude within yourself to start over again, this time it will be a very different scenario.

Unlike your first year, when you started out with a pristine, clean, empty hive and brand-new blank top bars, now you have something known as drawn comb. It represents a pretty hefty amount of effort by your former bees, which your new bees will not have to expend. It's not that they aren't amazingly good at making wax, but a new package of bees is a fairly fragile beginning for a hive, and having some bars of already drawn wax will benefit them twofold. One, they have drawn wax with empty cells, and the queen can begin laying eggs the moment she is released from the queen cage, because the empty cells are there, ready and waiting. And two, you no longer need a *starter kit* to *anchor* your bees into an empty top bar hive. You've got something much better than a starter kit—you've got fully drawn combs!

Installing Bees in Year Two with Drawn Comb

Now that you have a hive with drawn comb, how do you configure the hive when preparing to install new bees? Last year you installed your bees into the space of 10 or so empty top bars. This year, you probably have 10 or more bars of drawn comb—so where should you put it all? These are questions relevant to your second season, so let's talk now about configuring your top bar hive to start again in year two, with drawn comb and new bees.

The typical top bar hive started from a package usually begins with the beekeeper installing the bees into a *bee bowl*—a space of approximately 10 bars, perhaps a few less if the weather was still cold when your bees were hived, perhaps a few more if it was later in the season or warm. It's a bigger challenge to hive bees in very cold weather in a brand-new empty top bar hive. Without drawn comb, the bees have nothing to climb on up to the top bars to form the cluster they must create to stay warm and to survive; they can only clamber up the inside of the hive. But with comb to use as a ladder, this scenario improves.

So let's suppose that you can start bees in this second-year hive, with 10 bars or more of drawn comb to go into what last we called the bee bowl. Set up your follower board or boards as you did for last year. The bars you remove in order to install the bees into the hive will now have comb on them. Remove these bars from the hive and set them aside somewhere safe, perhaps inside the overturned lid of the hive, lying on the ground next to you while you are preparing to hive the bees. Be careful to preserve the order and direction of each of the bars.

Remove the lid from the bee package, and investigate the configuration of the package itself. Some packages require that the syrup feeder can be removed first; some allow the queen cage to be removed before the can. In any event, work these two items out of the package and leave it loosely covered with the lid to contain the bees while you get started.

To hang the queen cage, choose either a blank bar or a bar with only a small piece of comb hanging from it, which leaves room enough to attach the queen cage. Remove the cork that covers the "slow-release" candy plug and hang her from the bar you've chosen, placing this bar in the number 1 position in the brood nest.

Bonk and gently pour the rest of the bees into the bee bowl. So far this has been very much like your first year, hasn't it? Except now you want to put the bars back into the hive, and there is comb on them! This complicates things a little bit, but it's definitely a good problem to have. Slowly lower each bar into place, one at a time, moving bees gently off the top edge of the hive as you go, until all the bars are back in the hive. The last one will be the most difficult to insert since it's no longer just an empty bar to drop into place but now has comb attached, but no worries. Here's the trick: Just remove 1 or 2 bars from the hive on the far side of the follower board, and move the follower board over to get the room you need. When that last bar of comb is in place, put the follower board back where it belongs, and then reinstall the blank bar(s).

Just like last year, check in 3 to 5 days to see that your queen has been released. If she has, remove the queen cage. If she has not, you should release her. Remove the queen cage from the bar, and holding the cage deep down in the hive, remove the staple that holds the screen, letting the queen walk out into the hive.

The rest of the year's buildup and maintenance will mimic the hive's first year, except that the queen will already have places to lay eggs and the hive will be able to build up the brood nest much faster than in your first year.

It's Year Two and You Don't Know Yet If You've Still Got Bees...

This is a long-time conundrum in beekeeping. If you buy a package thinking you are sure to need it, but then your bees make it through the winter, now you've got more bees than you need, possibly even more bees than you have hives. But if you don't buy a package and your bees don't make it through the winter, now you've got fewer bees than you wanted. And of course, there's a lot of pressure to order your bees early; bee suppliers often sell out in mid-spring, even before you can get to your bees to check on them, depending on where you live and what your winters are like.

One way to deal with this dilemma is a "just-in-case" package of bees. Gold Star Honeybees instituted a practice to deal specifically with this very situation. It works like this: It's early spring, and it looks like your bees are still going strong—but April is not over yet, so you may still lose them.

In a "just-in-case" scenario, you order a package just as you normally would, but if your bees are still alive, and you don't need the new package, you can cancel the purchase (in a timely manner!) and pay only a nominal cancellation fee.

This way you don't have to worry about missing out on bees for an entire year if your bees don't overwinter, and you also don't have to worry about having more bees than you planned on when your bees do successfully overwinter. Pretty good insurance, just in case, isn't it?

part 2

Smoke and Mirrors

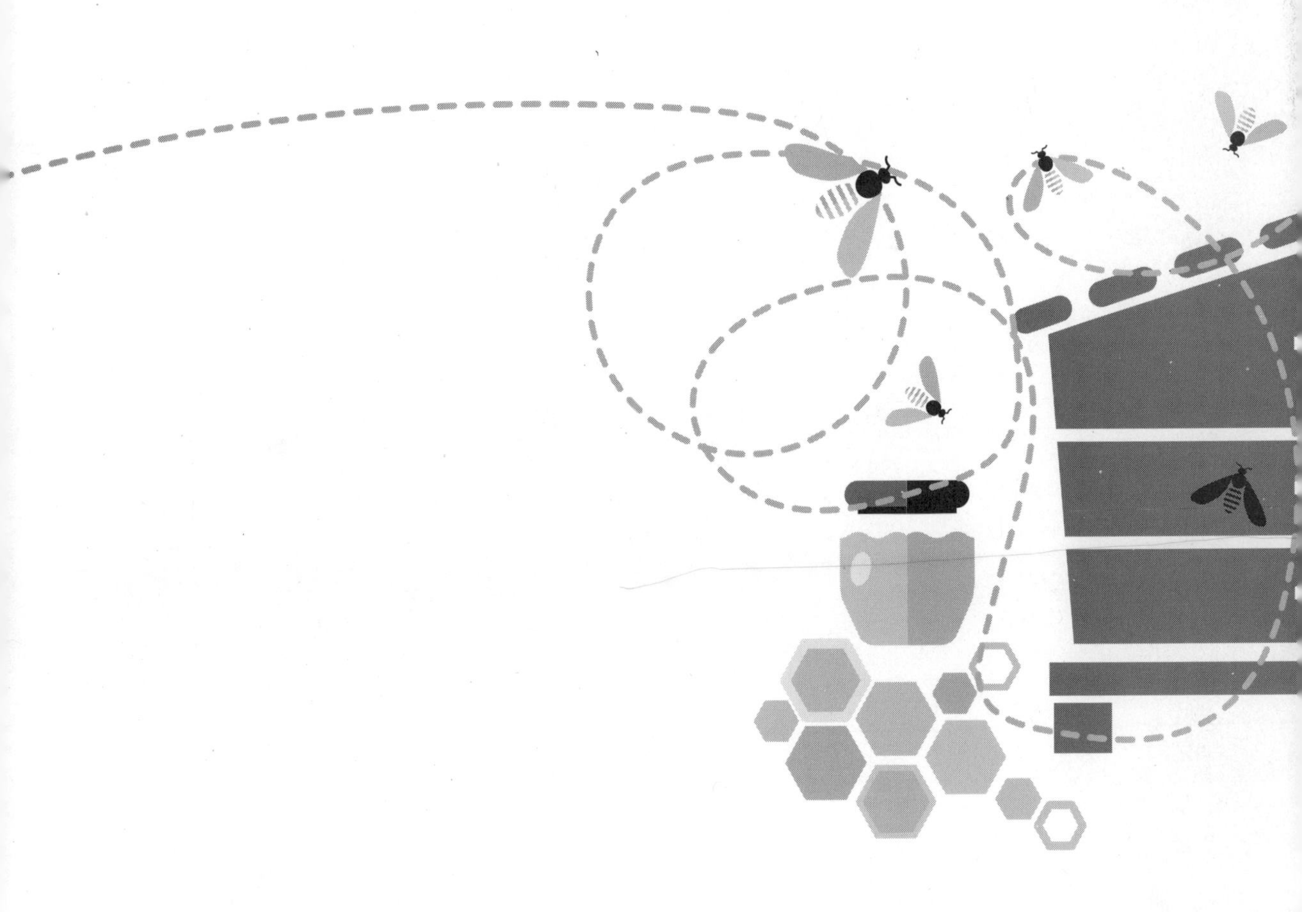

8 What Is Really Going on with Bees? *or* Is It Time to Redefine the Crisis Vocabulary?

In November 2006, a beekeeper by the name of David Hackenberg was having a problem with his bees. Hundreds of his hives went from thriving colonies to empty beehives over the course of a few weeks. Disturbed by his own bee troubles and bolstered by the surreptitious rumblings of other beekeepers who were experiencing similar problems, Dave went to the bee researchers at Penn State University for help. It wasn't long before Dave Hackenberg was considered the poster child for a frightening problem threatening honeybees—soon to be named *Colony Collapse Disorder* or CCD. In 2008 he received the American Beekeeper Federation's President's Award for highlighting this problem, and he has testified before the US Congress and prominent international groups concerning pesticides and bees.

That was a decade ago. Where are we with this "bee problem" today?

Colony Collapse Disorder (CCD)

CCD was originally thought to be a single problem—a disease or disorder—with a single cause. Research money was devoted to studying the phenomenon, and trying to narrow down the field of possibilities to ferret out that one cause. Collapsing hives were studied in search of a common *marker,* one key indicator that would be found in all hives said to have

CCD. The search was on for a single solution to a single problem, in the face of a national panic over losing the honeybee altogether. That's the way our science has generally worked, at least in the past. One problem should have one cause. It simplifies the science when things work that way, and makes it easier to do research. If you can identify that one cause, then you can resolve that one problem. Voilà!

But it wasn't long before the beekeeping research community realized that there was no single cause of CCD. That in fact, it was more than a little naïve to view beekeeping and its attendant issues in such simplistic terms. More than one prominent researcher has put it this way: CCD is caused by a combination of things, and that is frightening, because the number of possible combinations is infinite, and unknowable.

Since 2006, CCD has become more of an umbrella term than the name of a specific condition, and it is covering a lot of ground. It represents an attempt to recognize all the known influences, the stressors and the many diverse factors affecting honeybees today—and to group them all under a common name.

Perhaps it would be more accurate to describe the problems with bees as *Nature Deficit Disorder*, a phrase Richard Louv coined in his 2005 book *Last Child in the Woods*. When describing the effects that our disconnectedness from nature is having on honeybees, it seems an appropriate moniker.

So just what are those things that are weighing so heavily in combination on the tiny honeybee? Here are seven things, like as not an incomplete list, but that will serve as a starting place:

Varroa Mites

This pest arrived in the US in the mid-1980s and now troubles beekeepers everywhere. Our first response to the realization that this mite had jumped species from an Asian bee—an insect fairly well equipped to deal with it—to the European honeybee, which was not, was chemical warfare, in the hope of eradicating the now ubiquitous *varroa destructor*.

Enter the first of two highly toxic and fat-soluble miticides: *coumaphos*. Beekeepers took a zero-tolerance stance toward this parasite and went after it with guns a-blazing—placing plastic strips containing this toxic chemical directly into the brood nest of the beehive.

When the mites developed resistance to coumaphos, the industry responded with a second chemical, the synthetic pyrethroid *fluvalinate*. It wasn't long before varroa developed a resistance to this toxic treatment as well.

Few studies had been done on the long-term effects on the honeybee from these chemicals individually...and now there was a cocktail of chemicals in the hive, and worse, they were also in the wax.

Contaminated Wax Foundation

Once the problems with mites began, and toxic chemicals were deployed inside beehives in an attempt to eliminate the mites, another important scientific truth made itself apparent. Beeswax is lipophilic. Lipophilic, in layman's terms, means fat-loving, or absorbing fats.

What does this mean to bees? It means that the beeswax comb that they create inside their hive—both the brood combs, where they raise their young, and the honey combs, where they store their winter food supply of honey—have all been absorbing the toxic fat-soluble chemicals that beekeepers have been using in their hives for decades, in an unsuccessful attempt to control varroa mites.

Since the treating of bees and their hives for varroa mites began in the mid-1980s, combs that have been contaminated by these toxic chemicals have eventually been culled from the hive, rendered and remade into fresh wax foundation sheets.

But unfortunately, these chemicals are persistent pesticides. They are not eliminated when the wax is melted down to make new foundation. They do not evaporate or dissipate even when heated to the liquid state. The end result has been that the world's supply of beeswax foundation is now polluted with poisonous chemicals in readily detectable amounts. Even when brand-new, wax foundation is contaminated.

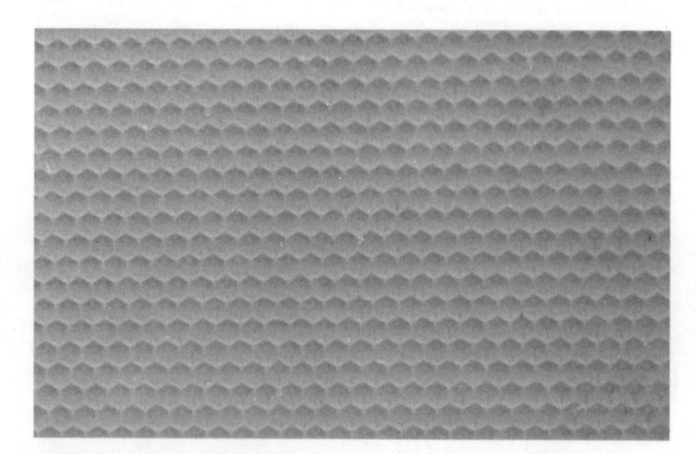

FIGURE 8.1. **Foundation determines the size, and thus, the gender of the bee.** Credit: Christy Hemenway.

Cell Size of Prefabricated Foundation

When beekeepers got the notion that they could save the bees a lot of work by using sheets of foundation, they went about it with a will, pressing out preprinted sheets of wax to give bees a head start.

The size of the brood cells made by the worker bees informs the queen which type of bee is to be raised in what cell and whether or not to fertilize the egg she is about to lay. A fertilized egg develops into a worker bee; an unfertilized egg becomes a drone. A change in the size of the cells that make up brood comb affects not only the size of the bee, but also the gender or *caste* of the bee.

Making the hexagons embossed on sheets of foundation an appropriate size for worker brood encouraged the raising of more worker bees, at the same time, it effectively prevented the colony from raising drone bees. This probably seemed like an ingenious idea at the time; after all, with more worker bees, there should be more honey available to harvest. The long-term effects of thwarting the bees' ability to raise the male of their species were slow to reveal themselves. This practice has limited the genetic diversity of the honeybee population in ways we are still discovering today.

Another important factor related to the use of foundation pertains to the size of the brood cells that all young bees—both workers and drones—develop within. The larger the brood cell, the larger the bee; the larger the bee, the longer its gestation cycle. When the life cycle of the honeybee is superimposed over the life cycle of the varroa mite, it becomes obvious that anything that extends the larval stage or the pupal stage of the honeybee's life cycle increases the success rate of the varroa mite, which breeds inside those same cells, along with the baby bees.

Homeowner Use of Chemical Pesticides and Herbicides

One area where chemical use has increased substantially is around the average suburban American house. The EPA permits more than 200 different pesticides for lawn care alone, and the US Fish and Wildlife Service reports that "homeowners use up to 10 times more chemical pesticides per acre on their lawns than farmers use on crops." This is not good news for the bees! Nearly 80 million pounds of pesticides are used on US lawns every year. That's not good news for birds either—as lawn care pesticides are among the most common causes of bird death.

Monoculture Agriculture

The industrial scale of American agriculture has played a large part in disrupting the balance and diversity of our food system. By its very nature, a huge *monoculture* farm, growing acre upon acre of a single crop, invites pests that thrive on that crop to move right in. At the same time, it forces out the pests that would otherwise be found there, and would work to maintain a natural balance, helping to mitigate the success of the original pests.

Because the natural predators of the invading pests have been eliminated, our very efficient industrial farmer now decides that the only course of action is to apply even more pesticides intended to control the original pest.

Pesticides, however, are dumb. They cannot tell the good bugs from the bad; so with this practice, every insect, beneficial or undesirable, important or insignificant, boll weevil or honeybee, is affected.

Migratory Pollination

When we adopted farming methods that were inspired by the early-1970s governmental directive to "Get big or get out!" the building blocks of large-scale American agriculture became the monoculture farmer. We wanted more "efficiency of scale," and we got it by first reducing the diversity of what we grew and then by growing enormous quantities of it, all in one place.

This misguided attempt to improve upon nature's processes destroyed the balance essential to small-scale diverse farming. It created an environment that works against the natural process of growing good food, and actually destroys the paradigm it was supposed to improve.

When we began growing food in this way, we ignored the basic needs of the pollinators that the plants required in order to produce their crop. In a monoculture environment, honeybees cannot survive naturally year round, because there is no forage for them once the bloom of the mono crop ends.

Monoculture creates the need to "bring the bees to the trees." We have begun to treat bee colonies as if they were just another piece of farming equipment. We burn huge amounts of fossil fuel to move hives to fields where they are needed, and then to move them out again. Meanwhile, the effects on the bees from being trapped inside their hives and moved long distances are still not well understood.

Destruction of Natural Bee Forage and Habitat

The urban destruction of natural fields, forests, and green areas limits the growth and health of the honeybee population by virtue of the simple fact that they have fewer places to live and less healthy, nutritious forage to consume.

These issues are just the tip of the iceberg when it comes to enumerating the damage that modern agriculture, toxic chemicals and urban sprawl have inflicted on the honeybee. Joni Mitchell described it succinctly in her song "Big Yellow Taxi"—"They paved paradise and put up a parking lot."

9 Are We Really Still Putting Chemicals in the Beehive?

SINCE THE MID-1980S, the world of beekeeping has really gone through some changes. Before then, it was pretty simple: get a beehive, get some bees, put the bees in the hive, watch them for the summer, harvest their honey, feed sugar if needed. Next spring: repeat this process, with the same bees. Split them as they prepare to swarm, and increase the size of your apiary at will.

With the advent of the varroa mite, this idyllic picture changed drastically. The use of miticides in beehives became the norm, and "when to use which chemical" became the lesson taught in beekeeping courses everywhere. Your granddad probably wouldn't recognize the way we keep bees today.

Mightier Mites

All of this new beekeeping education was well-intentioned, of course. Beekeepers do not want to lose their bees, and if killing the mites with a simple chemical worked, well then...why not? But the shortsightedness of this attitude became evident when the early chemical treatments began to lose their effectiveness. The pests they were meant to fight evolved and became resistant to the active ingredients in the treatments.

Resistance and Synergistic Effects

This resistance became the driving force behind the search for a new and more effective chemical to ward off pests. When one was found, we said, "Yay! A treatment that actually works!" Until once again, the pests developed resistance and were able to withstand the new treatments, and their effectiveness was diminished as well.

Now we had a new and different problem: the *synergistic effects* of multiple miticides. Synergistic effects are described as: "The interaction of two or more substances to produce a combined effect greater than the sum of their separate effects." We see this issue in other areas as well. The one that comes most readily to mind is pharmaceuticals. Remember how the last time you presented a prescription at a pharmacy and were asked straightaway by the pharmacist, "What else are you taking?" The concern that drives that question is real—it's about synergistic effects. There is a real possibility that otherwise helpful drugs could interact with each other in a way that could cause a serious health problem, or even be fatal.

But we don't ask the bees, "What else are you taking?" or the equivalent: "What else has been put into your hive?" The synergistic effects of putting one treatment after another into a beehive are impossible to quantify. The specific effects of individual treatments when used alone are one thing, but the potential reactions to the many possible combinations of chemicals and the differing quantities, timing and toxicities are myriad and impossible to track.

Persistent Pesticides in the Bees' Wax

When miticides were first developed and registered for use in beehives to combat varroa mites, the typical testing seemed to indicate that they were safe for use in the hive. The bees did not die immediately upon coming into contact with the treatments, but mites did, and the treatments were pronounced safe.

However, beeswax is lipophilic, meaning it absorbs fat-soluble substances such as miticides and other pesticides. As beekeepers continue to use toxic treatments in the hive, these treatments build up in that special stuff that the bees' nest is constructed of—their beeswax comb—the heart and the skeleton of their hive. The combs they build inside the beehive are used to store their honey and as a nursery, where eggs are laid and young bees are raised—two very special and important purposes!

Ultimately though, those same combs are eventually recycled and the wax is reused to make new sheets of foundation. Unfortunately, the pesticides we began using in the 1980s don't just evaporate when the wax is rendered...instead they persist, and even accumulate. Now even freshly made foundation wax is contaminated.

Weaker Bees

Bees don't die from their initial exposure to the active ingredients in products like CheckMite+® and Apistan®. But what have been the long-term effects of these products as they build up in the wax, stopping just short of killing the bees outright? Sure, the chemicals killed most, though not all, of the mites (whose survivors went on to develop resistance to this myriad of toxins)—but who was watching for the long-term cumulative effects on the bees?

Longevity and fecundity of queens; virility and fertility of drones; strength, health and resilience of worker bees; learning ability and memory; development of young bees through all the stages of their life cycle: from egg, to larva, to pupa, to bee; the vitality of the hive as a whole; the ability of bees to navigate from hive to sources of forage and to find their way back again—all of these have been affected since beekeepers began contaminating the bees' wax in the name of protecting the bees.

FIGURE 9.1. Plastic foundation in place of contaminated wax? Which is the lesser of the two evils here? Credit: Internet marketing photo.

Is Plastic Really Better?

One pragmatic response to discovering the contamination in beeswax foundation was to make foundation from something else, namely, plastic. This certainly eliminates the contaminated beeswax problem, but it leaves one wondering about how many new toxins are being introduced into the hive through off-gassing and other toxins frequently found in plastic.

The fact remains that the very nature of the preprinted embossed hexagons found on sheets of foundation—being all one size and slightly too large—whether constructed of plastic or wax, still makes it difficult for the colony to raise drones, and still adversely affects the gestation cycle of the worker bee, which supports the varroa mite's reproductive success.

Beekeeping Sans Foundation— Does Natural Wax Do Anything at All?

If one accepts the notion that an insect with 65 million years of practice at doing a thing probably knows how to do that thing quite well, then the idea that we humans, with our big brains and our opposable thumbs, could improve upon their process is preposterous. Yet the use of foundation began in the late 1800s, and its perceived benefits are extolled throughout the beekeeping industry.

The benefits of foundation seem to accrue primarily to the beekeeper, not necessarily to the bees, including: making it possible to harvest honey using an extractor, which subjects the comb and the honey to strong centrifugal forces in order to sling the honey from the comb; making hive inspections faster, by enforcing the creation of straight combs; and by making it easier for the beekeeper to handle the frames with less finesse while inspecting the hive.

If one further accepts the notion that the accumulation of persistent pesticides in beeswax foundation is a bad thing, then a beekeeping system that includes chemical treatments and uses foundation made from recycled contaminated beeswax is fraught with issues.

USDA AMS TESTED

United States Department of Agriculture | Agricultural Marketing Service | Science and Technology | National Science Laboratory 801 Summit Crossing Pl. Ste. B Gastonia, NC 28054

C. Hemenway

Applicant Identifier:
Maryann T. Frazier
Senior Extension Associate
Department of Entomology
501 ASI Building
University Park, PA 16802

Sample Description: **Wax**

Sample Identifier (Lab Code): **2010-026**

Internal Lab Reference Number: **AH39779**

Date Received: **8/26/2010 1**

Date Completed: **9/20/2010 12**

Pesticide Residue	Result (PPB)	LOD (PPB)	Pesticide Residue	Result (PPB)	LOD (PPB)
1-Naphthol	N.D.	10	Chlorpyrifos	N.D.	1
2,4 Dimethylaniline	N.D.	50	Chlorpyrifos methyl	N.D.	1
2,4 Dimethylphenyl formamide (DMPF)	N.D.	4	Clofentezine	N.D.	20
3-Hydroxycarbofuran	N.D.	4	Clothianidin	N.D.	1
4,4 dibromobenzophenone	N.D.	4	Coumaphos	N.D.	1
Acephate	N.D.	10	Coumaphos oxon	N.D.	1
Acetamiprid	N.D.	4	Cyfluthrin	N.D.	4
Acetochlor	N.D.	10	Cyhalothrin total	N.D.	1
Alachlor	N.D.	10	Cypermethrin	N.D.	4
Aldicarb	N.D.	4	Cyphenothrin	N.D.	20
Aldicarb sulfone	N.D.	3	Cyprodinil	N.D.	4
Aldicarb sulfoxide	N.D.	20	DDD p,p'	N.D.	20
Aldrin	N.D.	10	DDE p,p'	N.D.	2
Allethrin	N.D.	10	DDT p,p'	N.D.	2
Amicarbazone	N.D.	30	Deltamethrin	N.D.	20
Amitraz	N.D.	4	Diazinon	N.D.	1
Atrazine	N.D.	6	Dichlorobenzene-para	N.D.	10
Azinphos methyl	N.D.	6	Dichlorvos (DDVP)	N.D.	10
Azoxystrobin	N.D.	2	Dicofol	N.D.	1
Bendiocarb	N.D.	2	Dieldrin	N.D.	10
Benoxacor	N.D.	4	Difenoconazole	N.D.	10
BHC alpha	N.D.	4	Diflubenzuron	N.D.	20
Bifenazate	N.D.	20	Dimethenamid	N.D.	10
Bifenthrin	N.D.	1	Dimethoate	N.D.	20
Boscalid	N.D.	4	Dimethomorph	N.D.	20
Bromuconazole	N.D.	20	Dinotefuran	N.D.	30
Buprofezin	N.D.	20	Diphenamid	N.D.	1
Captan	N.D.	10	Endosulfan I	N.D.	2
Carbaryl	N.D.	3	Endosulfan II	N.D.	2
Carbendazim (MBC)	N.D.	5	Endosulfan sulfate	N.D.	2
Carbofuran	N.D.	1	Endrin	N.D.	10
Carboxin	N.D.	4	Epoxiconazole	N.D.	1
Carfentrazone ethyl	N.D.	1	Esfenvalerate	N.D.	2
Chlorfenopyr	N.D.	1	Ethion	N.D.	10
Chlorfenvinphos	N.D.	6	Ethofumesate	N.D.	5
Chlorferone	N.D.	50	Etoxazole	N.D.	1
Chlorothalonil	N.D.	1	Etridiazole	N.D.	10
Chlorpropham (CIPC)	N.D.	40	Famoxadone	N.D.	20

LOD - Limit of Detection, N.D. - Not Detected.

The fee for the laboratory services provided above is $273.00.

The U.S. Department of Agriculture (USDA) prohibits discrimination in all its programs and activities on the basis of race, color, national origin, age, disability, and where applicable, sex, marital status, familial status, parental status, religion, sexual orientation, genetic information, political beliefs, reprisal, or because all or part of an individual's income is derived from any public assistance program (Not all prohibited bases apply to all programs.) Persons with disabilities who require alternative means for communication of program information (i.e., Braille, large print, and audiotape) should contact USDA's TARGET Center at (202) 720-2600 (voice and TDD). To file a complaint of discrimination, write to USDA, Director, Office of Civil Rights, 1400 Independence Avenue, S.W., Washington, D.C. 20250-9410, or call (800) 795-3272 (voice) or (202) 720-6382 (TDD). USDA is an equal opportunity provider and employer.

Approved by: Roger Simonds

Roger Simonds, Laboratory Manager 9/20/10 **Page 1 of 2**

FIGURE 9.2a. These lab test results show the natural wax from a Gold Star top bar hive is clean! Credit: Christy Hemenway.

Pesticide Residue	Result (PPB)	LOD (PPB)	Pesticide Residue	Result (PPB)	LOD (PPB)
Fenamidone	N.D.	10	Potasan	N.D.	50
Fenbuconazole	N.D.	2	Prallethrin	N.D.	4
Fenhexamid	N.D.	6	Profenofos	N.D.	10
Fenoxaprop-ethyl	N.D.	6	Pronamide	N.D.	1
Fenpropathrin	N.D.	1	Propachlor	N.D.	10
Fenpyroximate	N.D.	5	Propanil	N.D.	10
Fipronil	N.D.	10	Propargite	N.D.	10
Flonicamid	N.D.	8	Propazine	N.D.	4
Fludioxonil	N.D.	20	Propetamphos	N.D.	4
Fluoxastrobin	N.D.	4	Propiconazole	N.D.	10
Fluridone	N.D.	10	Pymetrozine	N.D.	20
Flutolanil	N.D.	4	Pyraclostrobin	N.D.	15
Fluvalinate	N.D.	1	Pyrethrins	N.D.	50
Heptachlor	N.D.	4	Pyridaben	N.D.	1
Heptachlor epoxide	N.D.	10	Pyrimethanil	N.D.	3
Hexachlorobenzene (HCB)	N.D.	1	Pyriproxyfen	N.D.	1
Hexythiazox	N.D.	6	Quinoxyfen	N.D.	10
Hydroprene	N.D.	10	Quintozene (PCNB)	N.D.	1
Hydroxychlorothalonil	N.D.	50	Resmethrin total	N.D.	10
Imazalil	N.D.	5	Sethoxydim	N.D.	2
Imidacloprid	N.D.	1	Simazine	N.D.	10
Imidacloprid 5-hydroxy	N.D.	25	Spirodiclofen	N.D.	1
Imidacloprid olefin	N.D.	10	Spiromesifen	N.D.	10
Indoxacarb	N.D.	3	Tebuconazole	N.D.	8
Iprodione	N.D.	20	Tebufenozide	N.D.	5
Lindane	N.D.	4	Tebuthiuron	N.D.	2
Linuron	N.D.	20	Tefluthrin	N.D.	1
Malathion	N.D.	4	Tetrachlorvinphos	N.D.	4
Metalaxyl	N.D.	2	Tetraconazole	N.D.	6
Methamidophos	N.D.	4	Tetradifon	N.D.	1
Methidathion	N.D.	10	Tetramethrin	N.D.	10
Methomyl	N.D.	10	Thiabendazole	N.D.	1
Methoxyfenozide	N.D.	2	Thiacloprid	N.D.	1
Metolachlor	N.D.	6	Thiamethoxam	N.D.	4
Metribuzin	N.D.	1	THPI	N.D.	50
MGK-264	N.D.	10	Triadimefon	N.D.	2
MGK-326	N.D.	10	Triadimenol	N.D.	45
Myclobutanil	N.D.	5	Tribufos (DEF)	N.D.	2
Norflurazon	N.D.	6	Trifloxystrobin	N.D.	1
Oxamyl	N.D.	5	Triflumizole	N.D.	10
Oxyfluorfen	N.D.	1	Trifluralin	N.D.	1
Parathion methyl	N.D.	2	Triticonazole	N.D.	10
Pendimethalin	N.D.	6	Vinclozolin	N.D.	1
Permethrin total	N.D.	10			
Phenothrin	N.D.	10			
Phorate	N.D.	10			
Phosalone	N.D.	10			
Phosmet	N.D.	10			
Pinoxaden	N.D.	5			
Piperonyl butoxide	N.D.	6			
Pirimiphos methyl	N.D.	4			

LOD - Limit of Detection, N.D. - Not Detected.

The fee for the laboratory services provided above is $273.00.

The U.S. Department of Agriculture (USDA) prohibits discrimination in all its programs and activities on the basis of race, color, national origin, age, disability, and where applicable, sex, marital status, familial status, parental status, religion, sexual orientation, genetic information, political beliefs, reprisal, or because all or part of an individual's income is derived from any public assistance program (Not all prohibited bases apply to all programs.) Persons with disabilities who require alternative means for communication of program information (i.e., Braille, large print, and audiotape) should contact USDA's TARGET Center at (202) 720-2600 (voice and TDD). To file a complaint of discrimination, write to USDA, Director, Office of Civil Rights, 1400 Independence Avenue, S.W., Washington, D.C. 20250-9410, or call (800) 795-3272 (voice) or (202) 720-6382 (TDD). USDA is an equal opportunity provider and employer.

Approved by: Roger Simonds

Roger Simonds, Laboratory Manager 9/20/10 Page 2 of 2

FIGURE 9.2b. These lab test results show the natural wax from a Gold Star top bar hive is clean! Credit: Christy Hemenway.

So in the summer of 2010, when I got the opportunity to submit a sample of natural beeswax from a year-old, overwintered top bar hive from Maine to the lab that did the testing for a research project published by some great folks at Penn State, I was really excited. Was the wax going to be as clean as I hoped? I sent in the sample, as instructed, and a few weeks later, I received the lab results.

It took a moment to decipher them. All of the chemicals that were being tested for showed a result, in PPB, of N.D. What the heck was that, I wondered? I searched for the legend, and N.D., it turns out, stands for "Not Detected." Not detected? Really? Wow. *None* of the chemicals they were testing for were found in our natural wax sample.

We feel very strongly that the wax in a beehive should be clean, natural beeswax, made *by* bees, *for* bees... so those lab results were nothing short of vindicating. It really is "All About the Wax!"

In a nutshell, under Methodology/Principal findings, the authors stated:

> We have found 121 different pesticides and metabolites within 887 wax, pollen, bee and associated hive samples. Almost 60% of the 259 wax and 350 pollen samples contained at least one systemic pesticide, and over 47% had both in-hive acaricides fluvalinate and coumaphos, and chlorothalonil, a widely used fungicide.

And under Results, they stated: "Only one of the wax, three pollen and 12 bee samples had no detectable pesticides."
(journals.plos.org/plosone/article?id=10.1371%2Fjournal.pone.0009754)

10 Wait a Minute: Wasn't Nicotine a Bad Thing?

When human activities may lead to morally unacceptable harm that is scientifically plausible but uncertain, actions shall be taken to avoid or diminish that harm.

— The Precautionary Principle as enshrined in the EU Commission's directive 91/414, and defined by UNESCO in 2005.

The Precautionary Principle

It's 2016. We know cigarette smoking is bad for your health. We know it causes cancer. We also know that the tobacco companies deliberately deceived the American public about the use of tobacco for a very long time. They lied about the health effects of smoking on the smoker, and about the addictiveness of nicotine. They lied about the fact that cigarettes were intentionally designed to increase the nicotine delivered to the smoker, about the fact that "light" or "low tar" cigarettes are not less harmful than regular cigarettes and about the effects of secondhand smoke. The depth of the deception is... well, let's just say it's deep!

In November 2012, US District Court Judge Gladys Kessler set forth the text of the corrective statements that the tobacco companies were ordered to publish, to inform the public about the truth. The deadline for compliance was March 1, 2013, but the tobacco industry has yet to comply.

Just to be sure we grasp the scope of this issue, let's review the details of the Order that Judge Kessler sent back to the courts for action then and the actual statements the defendants in *United States of America v. Philip Morris USA, Inc., et al.* were ordered to make known (tobacco-on-trial.com/wp-content/uploaded/2012/11/20121127-doj-5991-order-_34-remand.pdf; care2.com/causes/5-lies-the-tobacco-companies-legally-have-to-take-back.html; care2.com/causes/supreme-court-declines-case-big-tobacco-scores-big.html; npic.orst.edu/factsheets/archive/imidacloprid.html).

Here is the truth about the *adverse health effects* of smoking:

- Smoking kills, on average, 1,200 Americans, every day.
- More people die every year from smoking than from murder, AIDS, suicide, drugs, car crashes and alcohol, combined.
- Smoking causes heart disease, emphysema, acute myeloid leukemia and cancer of the mouth, esophagus, larynx, lung, stomach, kidney, bladder and pancreas.
- Smoking also causes reduced fertility, low birth weight in newborns and cancer of the cervix and uterus.

Here is the truth about the *addictiveness of smoking and nicotine*:

- Smoking is highly addictive. Nicotine is the addictive drug in tobacco.
- Cigarette companies intentionally designed cigarettes with enough nicotine to create and sustain addiction.
- It's not easy to quit.
- When you smoke, the nicotine actually changes the brain—that's why quitting is so hard.

The tobacco companies *advertised "low tar" and "light" cigarettes as less harmful than regular cigarettes. They are not.* Here's the truth concerning that claim:

- Many smokers switch to low tar and light cigarettes rather than quitting because they think low tar and light cigarettes are less harmful. They are *not.*

- Low tar and filtered cigarette smokers inhale essentially the same amount of tar and nicotine as they would from regular cigarettes.
- *All* cigarettes cause cancer, lung disease, heart attacks and premature death—lights, low tar, ultra lights and naturals. There is no safe cigarette.

The tobacco companies *deliberately deceived the American public about designing cigarettes to enhance the delivery of nicotine.* Here's the truth:

- Defendant tobacco companies intentionally designed cigarettes to make them more addictive.
- Cigarette companies control the impact and delivery of nicotine in many ways, including designing filters and selecting cigarette paper to maximize the ingestion of nicotine, adding ammonia to make the cigarette taste less harsh and controlling the physical and chemical makeup of the tobacco blend.
- When you smoke, the nicotine actually changes the brain—that's why quitting is so hard.

Then there are the devastating effects of *secondhand cigarette smoke.* Here is the truth:

- Secondhand smoke kills over 3,000 Americans each year.
- Secondhand smoke causes lung cancer and coronary heart disease in adults who do *not* smoke.
- Children exposed to secondhand smoke are at an increased risk for sudden infant death syndrome (SIDS), acute respiratory infections, ear problems, severe asthma and reduced lung function.
- There is no safe level of exposure to secondhand smoke.

We should be outraged. More than 20 million Americans have died since 1964 because of tobacco. Many of them (approximately 600,000) didn't even smoke cigarettes! According to the American Cancer Society, tobacco is the most preventable cause of death in the US; it is responsible for 30 percent of all cancer deaths and 87 percent of all lung cancer deaths. Yet the tobacco industry's subterfuge goes unpunished. Interesting...

Let's stay with this subject of nicotine for just a moment longer. Today's thinking beekeepers, and bee-friendly gardeners, are becoming aware of a new nicotine issue: a new class of pesticides that arrived on the scene in 1994—*neonicotinoid* insecticides. These *systemic pesticides* differ from others in that they permeate the tissue of the entire plant, rather than being found only on the exterior surfaces. The toxin is found *in* the seed and the roots, the stems and buds, the leaves, the flowers, the fruit—all of it, every cell, throughout the whole plant, pollen and nectar included! Surely this spells trouble for honeybees.

The most prevalent use of systemic neonicotinoid pesticides is as a prophylactic seed treatment on corn, soybeans, canola and sunflower seeds. Oddly, the United States Department of Agriculture does not even include systemic pesticides used as a seed treatment when they gather data and statistics on pesticide use, despite the fact that seed treatments account for approximately 95% of the total use of these pesticides.

Some folks say that these chemicals are harmful. Some say that they are safer than the older classes of pesticides. Some point out that if we are using systemic pesticides, then we are spraying fewer pesticides on our fields, which pollute the soil, the air and the water, and isn't that a good thing? Some people say that systemic neonicotinoid pesticides are the cause of Colony Collapse Disorder.

The chemical technology industry has a vested interest in controlling public perception of the safety of systemic pesticides, and they employ many resources—financial, marketing, and legal—in skewing that perception in their favor.

Don't these questions, and this subterfuge, sound familiar? If you returned to the beginning of this chapter and replaced tobacco with systemic neonicotinoid pesticides, wouldn't you find many frightening parallels between the two?

11

OMGs!! GMOs!! ABCDEs!! M-O-U-S-E!!

I'M GUESSING THAT the last chapter probably left you feeling that there's a lot of confusion and misinformation concerning the use of systemic neonicotinoid pesticides. The same comment could easily be made about the terms *genetically engineered* (GE) or *genetically modified organism* (GMO). First of all, there's a tremendous amount of confusion just about the use of these two terms. Do they mean the same thing or is there an accepted technical definition of each of them? That's a good question. So far, there doesn't seem to be a terribly good answer.

In fact, GMO and GE are so frequently used interchangeably that making a case for the differences between them feels a bit like splitting hairs. Do they really say anything different? I'd say yes, and so might some others, but remember that you are currently living in the world according to Christy Hemenway...so what I'm going to tell you is, of necessity, weighted heavily toward my own opinion, since it just didn't prove easy to find any reliable source that one could call "technically correct."

With that in mind, here's how I'd use the two terms: Genetically modified would tend to denote a plant or an animal whose genetics have been purposely guided over time using natural sexual reproductive processes. This could be likened to breeding roses for a particular color, or pumpkins for size or horses for speed. It happens by pairing members of a species that share a desirable trait with the intent of increasing the instances of that trait, and the trait is expressed in the offspring of those members. This process is the basis of the "survival of the fittest" mechanism, and over time its outcomes are seen as evolution.

"Genetically engineered" refers to a more industrial process that forcibly incorporates the genes of organisms that are not sexually compatible, using recombinant DNA techniques. Genetically engineered crops are purposely created, or engineered, by intentional manipulation of the genes of the organism. The resulting plants would not occur in nature without this highly technical manipulation. So genetic engineering is a form of genetic modification, but genetic modification is not necessarily genetic engineering.

Under the heading of genetic engineering, we find *Bacillus thuringiensis* or Bt crops. The primary argument in favor of these is that due to the plant's altered genes, which incorporate a soil bacterium that causes the plant to produce *Bt toxin* internally, we should be able to use fewer insecticides. Bt crops have been modified genetically so that the plant produces a toxin that kills pests by essentially blowing up their stomachs. However, the Bt toxin produced by the GE plant is much more concentrated than Bt sprayed topically, so in reality, the use of Bt pesticides has actually increased exponentially.

Agrochemical and agricultural biotechnology corporation Monsanto originally claimed that Bt would be destroyed as it traversed the digestive system of humans, and so could be said to pose no health risk. This was proven false when Bt toxin was found to bioaccumulate in the human body. Research suggests that Bt toxin may also cause immune responses that have been associated with allergies and infections.

And stop me if you've heard this one before—but it turns out that the pests in question have developed resistance to the Bt toxin. *Now instead of being able to use fewer pesticides, these resistant pests require the use of more and more pesticides in order to be effective.* Hmmm…

Other examples of genetic engineering are the herbicide-resistant or "Roundup Ready" soybeans, corn, cotton, canola and alfalfa. These plants have been modified to make them impervious to the effects of Monsanto's herbicide Roundup. This means the farmer can spray their field with it to kill weeds and unwanted plants, but the desired crop is unaffected. Gross, huh?

No wonder there is such a pitched legal battle occurring today concerning the labeling of such crops. If we were reminded to worry about the possible unintended consequences of genetic modification or genetic engineering every time we purchased a genetically engineered product—much like the tobacco industry was concerned would happen—surely we would want to quit!

FIGURE 11.1. Labels matter! Credit: Justlabelit.org.

Honeybees engender a deep passion among beekeepers. I would venture to say that there is no such thing as a beekeeper that does not love their bees. It is easy to become emotional about a subject like losing the honeybee, especially with the undercurrent of suspicion the public carries toward the actions of the pesticide industry. It can be difficult to have a thoughtful conversation with such impassioned believers on both sides of the issue.

Because I am a beekeeper, and because my "crunchy-granola" gene is bigger than my head, I have spent considerable reading time digging into the subjects of organics and GMOs and pesticides and cancer and vaccines and raw milk and corporate money and Monsanto and Bayer and Syngenta...only to quickly find myself drowning in the worst case of overwhelm since I first wanted to learn to keep bees in top bar hives and went looking on Google.

The internet is an important and influential force in the world we now live in. As a method of sharing information about things you like, what you're doing or whom you're doing it with, it is amazing. Blog posts and social media have created avenues of self-expression that have elevated op-ed to (what passes for) journalism. I know; I run over 50 sponsored Facebook groups—a global group, and a group for all 50 states, as well as for several countries—in order to create a venue for top bar hive beekeepers to connect. Use these to find others, but remember that sharing links on Facebook and becoming enraged by the comments left as criticisms is no way to have an intelligent conversation.

So...yes, it gets confusing. You're not the only one feeling that way. The only solace I can offer, and it's a thin one, is to encourage you to always "consider the source." An adamant impassioned beekeeper, while having a huge heart in the right place, may not be the best source of up-to-date information. Then again, some of the best vehicles for creating smoke screens and spreading misinformation are the very professionally produced blog posts and quasi-informational websites—backed by the very people who are marketing the products you most distrust. Seek out peer-reviewed sources and writers with credentials and motivations in the right place. These will take you further than social media sensationalism any day of the week.

12 The Shamrock

THROUGH ALL MY MANY YEARS of growing houseplants, I have held a special spot in my Irish heart for what I always heard referred to as a "shamrock." As a houseplant, it comes in green or purple, with pretty little 5-petaled flowers. It wakes up in the morning by opening its triangular leaves to the light, and it goes to sleep at night by folding itself back up. It tolerates incredible abuse—like forgetting to water, or being mowed down to the soil by curious and apparently hungry cats. It gets by in low- to medium-light conditions; it's equally happy in a pot on a window ledge or in the ground. It's just not picky. If you let it really dry out, all the leaves and flowers will die off completely, but if you water it, it starts to perk up and produce new growth almost immediately. The same thing happens after my cats have completely mowed it down—it just comes right back. It grows from the strangest-looking little tubers—they look and feel like little plastic pinecones. They grow longer over time and sometimes protrude above the surface of the soil. If you break them off, you can grow another plant from the pieces. It's such a cheerful and tenacious little plant—much more resilient than its fragile stems and thin leaves and delicate flowers would lead you to believe. I love the look of it, and its amazing resilience, and its adorable wake-up/go-to-bed behavior

FIGURE 12.1. A shamrock, source of oxalic acid.
Credit: Christy Hemenway.

just makes me smile. I bought my first one at a St. Patrick's Day sale at a local greenhouse, where I'm sure they called it a shamrock. So it took a little while before I learned about its actual background.

It turns out my sweet little shamrock plant is also known as wood sorrel. There are more than 800 different species, according to the *Encyclopedia Britannica*. They're found everywhere on the planet except for the poles. Sorrel is a word that comes from "sur" meaning sour. Its taste is quite tart or sour, much like lemon, and all of it—the leaves, the flowers, the roots or tubers—is edible. You can nibble on them or enjoy them in a salad. The tubers of some species are large enough that they are a popular vegetable, known as oca, or New Zealand yam. The tubers can be eaten raw, boiled or roasted. Eating too much wood sorrel could be a problem, as it has some calcium-leaching properties and can maybe affect your kidneys, but like most things that come with such a dire warning, you could probably never eat enough of it in one lifetime to be affected.

So by now, you'd probably like to ask me, "Christy, why have you just written almost 500 words on wood sorrel? Isn't this a beekeeping book?"

And indeed it is. And that's why, in the US in 2016, wood sorrel may be newly and differently relevant; the Latin name for this plant is oxalis. And oxalis is the source of *oxalic acid*, or OA.

OA has been registered as a pesticide in the past. At the time, its primary use was as a bathroom disinfectant. That registration was allowed to lapse in 1994. In 2015, it was again registered in the US—this time as a treatment for varroa mite control. Brushy Mountain Bee Farm is currently authorized by the EPA as the sole US distributor of oxalic acid as a varroa control. If you decide to use oxalic acid, be sure to read the label carefully and take advantage of the knowledge of the folks at Brushy Mountain, who can advise you as to its use.

Like most tired Americans, I am skeptical when I hear that a new drug or chemical is considered safe. Too many times we've run full speed ahead with a magical cure in the form of the most recent chemical, or pharmaceutical, or diet pill, only to discover too late that there were unintended consequences, trickle-down effects we hadn't considered, connections we hadn't bothered to honor, or even recognize. Some great tragedies come to mind in this realm—DDT, thalidomide, and DES, to name a few.

In some cases, the manufacturers of new drugs and chemicals purposely confuse the issue for their own personal gain. If you want a vivid example of this, recall the one in Chapter 10, which opened with a lengthy roster of untruths the tobacco companies told the American public about cigarette smoking and nicotine and secondhand smoke.

On the other hand, some fairly innocuous natural substances have been deemed dangerous and made illegal, based less on scientific evidence and more on the supposed detriment to society that they present. Marijuana is one example of this bias. Plenty of alarming things have been said about marijuana, but in this case, the lies set out to make the plant seem more virulent than it actually is.

I am a pretty serious zealot when it comes to raising bees without chemicals. I've been known as "the top bar whacko" since the start of my beekeeping career, so you know how much of a crunchy-granola tree-hugger that makes me. When I started keeping bees in 2007, the varroa mite had been a known issue for over 20 years, and Colony Collapse Disorder had been around just long enough to have been granted its own acronym, CCD.

But I also teach beekeeping, specifically beekeeping in top bar beehives. Not all, but many, of my students are brand-new to beekeeping, and they are very optimistic, and hopeful. And while they may be inexperienced, let me tell you—are they ever sincere. And they want nothing so much as to be able to keep their bees alive. Losing a hive of bees creates a heart-rending sense of personal failure.

This puts me in the interesting position of wanting to have good answers to their "new-bee" questions and good solutions to their beekeeping dilemmas. I look for solutions that fit into the treatment-free protocol that I advocate. So as well as I am able, I try to be objective and to consider all aspects of the problems faced by top bar bees and their keepers. I do my best to sort through the smoke and mirrors and the money-influenced politics and the brokenness of our agricultural system, and I search for solutions to issues that stand out above the "lesser of two evils" decision-making process we have adopted in these United States.

One of the most cherished side effects of having written the first book in *The Thinking Beekeeper* series was the *The Thinking Beekeeper* being translated into Italian in 2015. Not a lot of American beekeepers see much of what goes on outside the system of conventional beekeeping employed here in the United States. We are a very big country, and there's not much we need that we can't produce right here in our own great big backyard. But this means we sometimes don't look very far outside of our own culture, which can sometimes be detrimental. So when my first book was published in Italian, it presented me with the opportunity to check off an item that loomed large on my personal bucket list, and that was to travel to Italy, where I got the opportunity to talk with Italian beekeepers about keeping bees in top bar hives. (Of all things!)

Now, Italy's food system is a great deal different from ours. They don't use neonicotinoid seed treatments, and they don't believe in GMOs. Most of this stuff, common throughout the US, is going to be banned outright in much of Europe. And Italy has a cuisine that employs simple, natural, wholesome food prepared in simple, natural ways. They have organic certification mechanisms but they don't have to strongly differentiate good food from bad food, or safe food from toxic food. Everybody gets it that the food in Italy is great, right? It is. It's phenomenal. But I had no clue that it would be so relaxing. Relaxing? Yes, relaxing.

The level of suspicion relating to the production of food in America is pretty high. We suspect everything and everybody of not telling us the truth about our food—about what's in it, what's been done to it, how it's been grown. The current political unrest over a bill that's been nicknamed the DARK Act (Deny Americans the Right to Know) is about whether food companies should have to tell us when the food they are selling has been genetically modified. Not knowing for certain, but suspecting that you may have been misled about whether there is something harmful in the food you eat can be, well... stressful.

Italy's food system is less surreptitious, and less fraught with deception, and frankly, it's much more relaxing and a great deal less stressful. So the food really is one of the highlights of traveling to Italy, home to Carlo Petrini and the Slow Food movement. It's not only good food, as in pleasurable to eat, it's good food as in good for you, made of good things, made in good ways... just good.

Sounds heavenly, doesn't it? It is and I highly recommend it; if you get the chance to go to Italy, take it. But like American beekeepers, Italian beekeepers also have varroa mites. So when I got to visit bee yards in Italy, in the apiaries of entomologist Paolo Fontana, an experienced beekeeper and President of the World Biodiversity Association, I got to know more about oxalic acid, and how it's been being used successfully in Italy, and has been for years.

While I am definitely a crunchy-granola type, and a treatment-free top bar hive beekeeper, I am also very interested in non-toxic, effective ways of managing varroa mites. So in this chapter, purposely tucked away deep in the back of this book, we're going to consider the how and the why of using a chemical treatment for managing *varroa destructor*—the infamous varroa mite that has truly become the scourge of the beekeepers the world over.

Why to Use Oxalic Acid

There are three reasons why oxalic acid seemed like something a treatment-free beekeeper might consider using as a varroa control without having to completely abandon their partiality for being treatment-free. Firstly, there's the argument that oxalic acid is found in nature. It's even found in honey! Just like in the sweet little shamrocks growing in my kitchen, the level of naturally occurring oxalic acid is pretty low, so not harmful. The cats have certainly eaten plenty of shamrock and not died from it! Even the amount of oxalic acid found in the leaves of the rhubarb plant is small enough that you would have to consume a great quantity of it before you were seriously affected.

Secondly, remember how I'm always saying "It's all about the wax"? In that arena, oxalic acid has an important thing going for it—oxalic acid is not lipophilic. It seems likely that no oxalic acid residue builds up in the wax. In my opinion, this is a very important argument in favor of the use of oxalic acid when compared to other varroa control products.

Thirdly, thus far there seems to be no resistance to oxalic acid on the part of varroa. Resistance has occurred with other miticides in the past, and it would behoove us to remain skeptical and to stay alert for resistance issues with oxalic acid as well, but this does not seem to be happening. So that is another strong point in favor of oxalic.

When to Use Oxalic Acid

It's crucial to understand that, like most other varroa treatments, OA will only kill the phoretic mites, the ones that are out and about and riding on your bees; it will not do a thing to the mites that are tucked away inside the capped cells of developing brood. This means that its highest effectiveness is when little or no brood is present in the hive, so that all the mites are out of the cells and on the bees. The natural break in the brood cycle that occurs when your hive swarms can be utilized to meet this treatment requirement. Caging the queen for 18–24 days will

FIGURE 12.2. The dribble method of applying oxalic acid.
Credit: Randy Oliver.

also induce this broodless period, and the combination of the two is a common practice during the summer in Italy.

This technique, used after the main summer nectar flow has ended but before winter bees are being raised in the hive, offers several benefits. It lowers the mite load, lessens the field force of the hive and allows the winter bees to develop in an environment containing fewer mites.

How to Use Oxalic Acid

The how-to of using oxalic acid can be complex. It can be mixed into sugar syrup and applied directly to the bees as a measured dribble or as a spray, or the crystals can be sublimated. Using a special device to heat the OA, the acid sublimates, or turns into a vapor; the vapors then fill the hive and penetrate the cluster, killing the mites.

For a more detailed look at the history of its use as a miticide, I turned to Paolo Fontana. He has more than 30 years of beekeeping experience, and currently runs more than 70 top bar hives in Italy, in addition to his Dadant-Blatt hives, the Italian equivalent of the American Langstroth hive.

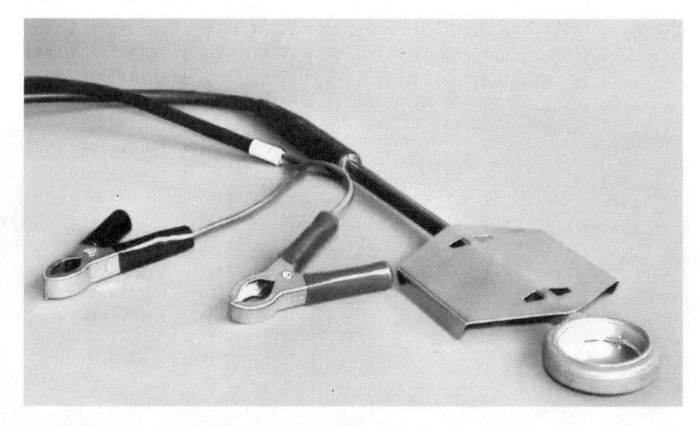

FIGURE 12.3. **The Varrox tool—used for sublimation method of applying oxalic acid.** Credit: Larry Welle.

Our discussion, presented below, addresses the details of how and when oxalic acid is being used in Italy—and illustrates the conversion of dosage calculations from the Dadant-Blatt hive to the BF (Biodiversity Friend) top bar hive that is rapidly catching on throughout Italy. And while he apologizes for what he calls his "macaroni English," I assure you that his English is far better than my Italian. (Another thing for the bucket list.)

Italian Experiences in Controlling Varroa Mites with Oxalic Acid

(by Paolo Fontana, edited by Christy Hemenway)

What I have written in these pages is the result of my personal practical experience in the apiaries of the Edmund Mach Foundation (San Michele all'Adige, Trento, Italy), where I coordinate a research group on bees and beekeeping, and also in the apiaries I manage with my friends (with which I founded a small society named "Apiamoci"). I have been a beekeeper for over 30 years. Furthermore, in recent years, I have had the opportunity to exchange experiences and opinions with many Italian and European beekeepers and researchers.

At the Edmund Mach Foundation, together with my colleagues and especially the researcher Valeria Malagnini, I have had the opportunity to assess on a large scale the advantages and disadvantages of different varroa

FIGURE 12.4. The first top bar apiary at Edmund Mach Foundation. Credit: Paolo Fontana.

control strategies. I have worked to convey a more rational, flexible, and efficient application of the biotechnical tactics that will be explained in full in the second part of this text. I also felt it would be useful specifically for top bar beekeepers to have a complete understanding of the different ways of using oxalic acid for controlling varroa mites.

The alternatives that I have presented can be defined as sustainable strategies. They are widely disseminated among Italian organic beekeepers but also even more by "conventional" beekeepers. These strategies are based on the biology both of honeybees and varroa mites and thus can be an intellectual resource for natural beekeeping with top bar hives.

The Use of Oxalic Acid to Control Varroa Mites in Italy

In 1982, after the introduction of the varroa mite, Varroa destructor, in Italy, Italian beekeepers began searching for control strategies that avoided the use of such chemicals as fluvalinate or amitraz. Many natural substances have been tested for their efficacy against varroa mites. Among these natural substances, there were three that gave good results: Thymol, an essential oil, and two organic acids, oxalic acid and formic acid. All of these products are permitted in Italy and throughout Europe in organic beekeeping.

Oxalic acid is an organic compound with the formula $C_2H_2O_4$, and it is naturally present, in small quantities, in honey. It is used by almost all Italian beekeepers, with different application methods and at different periods of the year. Damage to the bees caused by the application of oxalic acid varies depending on the mode of application and the frequency of its use in the hives. To date, there have been no confirmed cases of resistance to this organic acid in varroa mites.

Methods of Applying Oxalic Acid Treatment

Since 2006, only authorized veterinary products may be used in Italy. Today there is only one veterinary medicinal product containing oxalic acid and authorized for use in Italy for beekeeping, namely Api-Bioxal®.

Oxalic acid can be dispensed to the hive in three different modes. The simplest mode is dripping, or dribbling. The beekeeper dribbles between the combs a predetermined dose of a sugary solution of oxalic acid. For the dribbling method, 35 grams of Api-Bioxal® powder must be diluted in half a litre of a 1:1 sugar solution. It is applied using a calibrated medical syringe.

The second method consists of spraying the combs on both sides using a water solution of oxalic acid. To apply oxalic acid as a spray, 30 grams of Api-Bioxal® must be dissolved in 1 litre of water.

A third mode of use of oxalic acid is the sublimation of pure oxalic acid powder. The sublimated oxalic acid appears to have the fewest negative effects on bees. The dose for sublimation is 2.3 grams of oxalic acid powder per hive, to be inserted in the sublimator until complete evaporation.

Only Api-Bioxal® placed directly into the sublimator as powder is permitted. Sublimation requires specialized equipment that heats the oxalic acid powder up to its sublimation temperature, between 101° and 157° Celsius. Many of the tools available for sublimation of oxalic acid involve the use of electrical energy (batteries or connection to a car) or, less commonly, heat obtained from a gas flame. The oxalic acid vapour can be conveyed inside the hive by the main door or by openings made ad hoc in the back of the hives. The cost of the equipment used for sublimation may not justify their use in small apiaries.

Biotechnical Tactics against Varroa Mites Involving the Use of Oxalic Acid

Beekeepers tend to try to prevent their bees from swarming and, in the case of supersedure of the queen, tend to provide the colony immediately with a new fertilized queen. But in fact, it is known that honeybees use these two natural events to stop, even for a short period, the presence of brood, as well as to put a stop to the possible spread of brood diseases.

When a colony swarms, after about 20 days the colony will contain no capped brood. The swarming queen in fact normally stops laying eggs 3 to 5 days before swarming. The new queen will be born on the day of swarming or the next day and will start to lay eggs after 10 to 15 days; it will take other 8 or 9 days before any more capped brood is present in the colony. This chain of events mean that 20 days after swarming no capped brood will be present; this is the perfect time for a treatment with dribbled oxalic acid.

Italian beekeepers are the undisputed leaders (perhaps I brag a bit, sorry!) in the ingenious design and application of biotechnical tactics for the control of varroa mites. These different methods are united by one goal: to get, at a specific time, broodless colonies in which to use oxalic acid in a single application. This can eliminate 90% (or more) of varroa mites present in the whole colony. Oxalic acid works only on the phoretic mites—the ones running around loose on the backs of adult bees in the hive, not the ones reproducing inside the cells. It is completely ineffective on the mites sealed inside capped brood. Thus it is important for the hive to contain no capped brood when treating with oxalic acid. Purposely arranging for the absence of brood in the bee colonies during summer offers the chance for the bees to use their own biological strategy to contain brood diseases, as well as allowing the most efficient mode to control varroa mites. The biotechnical tactics can return to the bees the health benefits of swarming and supersedure and also provide an opportunity for the beekeeper to make a thorough check of the health status of their colonies.

Confinement of the Queen in the Cage or on a Single Honeycomb

A common technique among Italian beekeepers to get colonies without capped brood in the summer is queen confinement. This technique can be used on a large scale, by beekeepers with several hundred colonies. The crucial aspect of this technique is the requirement to find the queen.

There are many variations of this confinement technique, but the two main variables are the space in which the queen is confined and the duration of the confinement itself. With regard to the space, the queen can be confined on a honeycomb between two vertical queen excluders, or confined to a portion of a brood comb, allowing the queen to continue to lay eggs, although very limited, or it can be confined in a small cage with no possibility of laying eggs.

Something to Avoid

Many beekeepers tend to use oxalic acid incorrectly by neglecting to induce a broodless period but simply making repeated treatments in an attempt to catch the majority of the mites. This practice of numerous and repeated close treatments is harmful to bees, especially in the cases of dripped and sprayed oxalic acid application, but it is particularly so for the beekeeper when sublimation is adopted. The sublimation produces microcrystals of oxalic acid that are very harmful if inhaled, and if they come in contact with the ocular and oral mucous membranes. The use of sublimators must always be matched to that of masks with effective filters for fine particles (in Italy A3P3) and of appropriate protection for the eyes. Unfortunately oxalic microcrystals remain in the hive for a long time even after the use of the sublimator and can continue to be a danger to the beekeeper's health while visiting the hives in subsequent periods.

Oxalic Acid in the Control of Varroa Mites Specifically in Top Bar Hives

I have been a top bar beekeeper for only 2 of my 30 years as a beekeeper. I must say that it is a fascinating experience, and I would not have believed that the top bar hive works so well. In 2015, I populated and managed, in different apiaries, about 20 top bar hives, and by 2016, this number has exceeded 70 units. I am also in touch with dozens and dozens of new top bar beekeepers, so I constantly receive all their questions, and hear their problems, and their successes. There are no problems in applying most of the biotechnical tactics in a top bar hive in order to make the hive broodless.

My experience with the use of oxalic acid in top bar hives has been with the dripped or dribbled method. There are no particular contra-indications for use in top bar hives of oxalic acid sprayed and sublimated, although personally I do not support the spraying of oxalic acid as it is very disturbing for the bees, and I believe that sublimation makes us slaves of a yet another gimmick ready to malfunction or that requires energy in order to operate.

Dosage Calculations for Top Bar Hives

The other aspect that must be well verified is the calculation of oxalic acid dose to be applied. In Italy, in Dadant-Blatt hives the dose is of 5 ml for each space between the combs populated by bees both for dripped as well sprayed applications. To assess the dose for top bar hives the real comb area in the top bar hives must be calculated, and after that a calculation must be done of the surface area in other hives, and then on the dose used in that latter hives.

For example, the BF top bar, which I have developed in Italy together with some friends under the *Bees for Biodiversity* project, has honeycombs with an area of about 467 cm^2 that is approximately 40% of the surface of the typical Dadant-Blatt hive's combs (about 1152 cm^2); so the dose for dripped oxalic acid will be 2 ml (40% of 5 ml) for each space between the combs populated by bees. As discussed before, the 2 ml will be in a

FIGURE 12.5. **The size of a BF top bar hive fully built comb.** Credit: Paolo Fontana.

FIGURE 12.6. **The Chinese queen cage used to confine the queen. Workers can enter and exit this cage at will.** Credit: Paolo Fontana.

sugar solution for dripped application and in a water solution for sprayed application.

For the use of sublimated oxalic acid, the internal volume of the top bar hive must be calculated. If there are diaphragms or follower boards, only the volume containing combs between them must be considered. In this case, to find the right dose, a proportion with other hives must be calculated. For example, the BF top bar has a total volume of about 57 litres—which is about the same of a Dadant-Blatt hive (55 litres)—so to sublimate a hive full of combs (22 to 24 combs), 2.3 grams of oxalic acid powder must be used in the sublimator. If the BF top bar colony has only 12 combs within the diaphragms, then only 1.15 grams must be used.

Timing of Oxalic Acid Treatments

In Italy, from 2014 to 2016, following the *Bees for Biodiversity* project, we have gone from a situation where perhaps a few dozen colonies of bees were kept in top bar beehives to several hundreds. There are probably about 700 to 800 present nowadays, most of them in BF top bar hives. In most of the Italian top bar colonies, three treatments with dripped oxalic acid are done against varroa mites.

The first treatment is recommended and widely performed 7 or 8 days after introduction of the bees in the top bar hive, regardless that the hive is populated with a natural swarm or a bees package or transferring the queen and most of the bees from another Dadant-Blatt hive, previously situated in the place where the top bar will be allocated. This preventive treatment is very important because it postpones the moment when a significant population of varroa mites will be in the colony. If a top bar hive is populated at the end of spring or in early summer, this treatment may make unnecessary the summer treatment.

The second, and most popular, summer treatment in the Italian top bar hives involves the confinement of the queen. The queen can be captured and confined in what is known as the "Scalvini cage" or in the cheaper "Chinese queen cage"—types of queen cages that can allow to workers bees to move in and out (not the type used in packaged bees).

FIGURE 12.7. Using a queen excluder over a window in a thin diaphragm in a BF top bar hive.
Credit: Paolo Fontana.

This cage can then be hung on a top bar with no comb that is placed in the middle of the brood nest.

In several Italian top bar hives, we also successfully tested the confinement of the queen on a comb, enclosing it between two thin diaphragms (similar to a follower board but quite thin) having a large window made of a queen excluder. I am not much in favour of this last solution because it does not produce a total absence of brood for health purposes and because the queen may escape around the edges of the diaphragms, which must be attached securely to the top, bottom, and sides of the hive.

The third treatment in top bar hives is the winter treatment. The timing of this treatment can be verified checking the colony status. We can define the best moment when we find that in our colonies there are no more eggs but only open or capped brood. From the day we don't find eggs and young larvae in the hive, we can easily calculate that after two weeks the colony will be broodless. Sometimes in Italy, especially in the north, we observe broodless colonies in October, but in November they have brood, and then they are broodless again in December.

The biggest problem observed to date in the confinement of the queen in top bar hives is that the queens can escape from cages (though with

FIGURE 12.8. **Paolo Fontana among the Edmund Mach Foundation top bar hives.** Credit: Mirco Brugnara.

a low incidence), while the death of the queens and the lack of recovery of the egg laying seems less frequent than when used in standard hives. These considerations will be better evaluated over the coming years.

With regard to the length of the queen's confinement period, in the top bar hives we also recommend a duration of 24 days, which can be shortened to 18 days in less favourable climatic and environmental situations. The oxalic acid application must be done in both cases after 24 days. Every beekeeper must find his way in harmony with his own bees and take into account the local climate and environment.

Afterword

WELCOME TO THE END OF *The Thinking Beekeeper Thinks Again.* Oh wait…that's not what we called it in the end, is it? Well, if you've made it to here, we're going to assume that you know the name of the book you're holding in your hands…and by this time, you probably know what the book is about, and you probably know how important I believe it is that you understand the *why* of doing things. Of all things generally, but particularly when it comes to supporting natural systems, and even more specifically, when it comes to raising honeybees.

I am glad that you are here. I am proud and happy to be a part of your beekeeping journey. I also feel compelled to point out that 2016 is an election year in the US. It seems I suffer from an affliction that causes me to write beekeeping books during American election years. (I am so sorry!) But in this particular election year, with the extremism of the media circus involved in this particular American electoral process, I find that mostly…I am tired. I am tired of rhetoric, tired of disingenuousness and insincerity, tired of people saying one thing while they mean something else, and then actually do yet another, entirely different, thing. I am tired of politicians flip-flopping in pursuit of the political popularity bestowed by claiming to hold one opinion versus another. I am tired of the smoke and mirrors.

It's not just politics that makes me tired though; it's also money. It's money in politics. It's money in Big Agriculture and in Big Pharma. It's lots of money used to influence people away from doing the right thing and not enough money used to support a system of government once said to be "of the people, by the people, and for the people." It's lots of money being used to influence political decisions—even when those decisions become laws and policies that negatively affect our food supply, our environment, our planet; when they become laws and policies that affect the health and safety of our children, and our children's children.

It's disheartening. That's the first word that comes to my mind when I think of our views on food here in the United States. A trip to a grocery store can require a Sherlock Holmesian magnifying glass in order to suss out the truth of just what is in the package you are holding as you squint at the fine print on the label.

And how often does that label provide any information about how or where its contents were produced? Can you follow that chain back to its source? What are their policies about acceptable, sustainable growing methods? Where did they grow or process or package this food? Who profits when I lay my dollar down in exchange for this packaging-intensive food item? It needs to become simpler for the consumer, not more complicated. And yet we have things like the DARK Act, seemingly designed to keep us confused and, literally, in the dark about our food.

The suspicion and skepticism we have learned to harbor concerning our food supply does not nourish our bodies, our hearts, or our souls. It does not create a sense of safety, peace, or security when sitting down to eat together. Could this be part of why so many of us are experiencing disease, becoming obese, diabetic, cancer-ridden?

We are so accustomed to being lied to in this country. We expect corporate subterfuge and obfuscation. When we can battle our way through the smoke screen, we find outrageous policies that make no sense—even some that actually harm us. Sometimes this is enough to get us up on our feet to become activists: we write petitions; we work to put laws in place to protect our food, our children, and our planet. But sometimes it's all just too much. We are jaded and tired and bad-tempered—about *food*!

Food should nourish and support us and be a source of joy and satisfaction. When every simple question runs headlong into a brick wall, very little light can shine on our eating. Very little joy can grow from such a frustrated existence.

Our Environmental Protection Agency has become another political construct that we have learned cannot be trusted. The basic concept of checks and balances it was intended to maintain has crumpled beneath the weight of the almighty dollar in our agricultural industry. To set a fox to guard a henhouse is to invite that fox to a free meal. The results of testing done by industry-sponsored researchers smack more of marketing-speak and advertising than they do of clear, accurate information about your food.

As a nation, I think that nowadays we are all pretty tired.

So let's raise our sights a little. In early 2015, I was excited to learn that an independent group of scientists came together as the Task Force on Systemic Pesticides (TFSP). Their mission was "to carry out a comprehensive, objective, scientific review and assessment of the impact of systemic pesticides on biodiversity."

This multidisciplinary group consisted of 30 scientists from all over the globe. They synthesized 1,121 peer-reviewed studies published over the course of the previous five years, including some that were industry-sponsored. All of the TFSP publications were subject to the standard scientific peer review procedures of the Springer journal and were published in a special issue of *Environmental Science and Pollution Research* called "Worldwide Integrated Assessment of the Impacts of Systemic Pesticides on Biodiversity and Ecosystems." Eight scientific papers were produced. In summary, this work has shown "that the current large-scale prophylactic use of systemic insecticides is having significant unintended negative ecological consequences." [M. B. van Lexmond, J. Bonmatin, D. Goulson, D. A. Noome, January 2015, 22 (1), pp. 1–4]

Now that's not cheerful information—but at least it's clear. It's not sensational journalism, of which we have an overabundance during election years. It's not marketing spin by the corporations that manufacture systemic pesticides. It recognizes that there's a problem, that all-important first step in resolving any problem. So while it's bad news, it may also be the beginning of a move in the right direction.

I for one choose to see it as hopeful. The fact that so many independent scientists were concerned enough to come together on this project speaks volumes. It's not just all about chemical companies having their way—people do care. And from that caring place, let us all work together to do better—to take better care of our planet, ourselves, and…the bees.

Glossary

acaricides: See *miticides*

Africanized: honeybees with African genetics; usually quite aggressive and a challenge to work with

anchor: the top bar in a new top bar hive where the queen cage is hung; the bees will begin to build their comb here

bait and switch: the act of switching one hive for another; typically used to balance the population of a weak hive by placing it where a stronger hive had been in order to collect some of the strong hive's foragers

bearding: a large ball or cluster of bees hanging from the hive entrance, usually seen when the weather is extremely hot and/or humid

bee bowl: the open space in the hive where bees are poured when hiving

Big Ag: a colloquial term for policies, philosophies, and practices in industrial agriculture

bivouac: an interim location, in this case, the place where a swarm lands and stays until the choice is made as to the location of their new home

bloom: the plants in bloom at any particular moment in a specific locale, from which the bees are collecting nectar

bonk: an amusing term for the process of thumping a package of honeybees to knock them loose inside the package in preparation for hiving

break in the brood cycle: a period of time during which the hive contains no brood

bridge: bees are said to be bridging when they hang from a top bar in a catenary curve as they secrete the beeswax from which they construct comb; see *festoon*

brood comb: wax combs where brood is raised in a beehive

brood nest: the combs that house brood in the brood portion of the hive

Bt toxin: a bacteria that damages cell membranes, kills insects by affecting their gut; present in every cell of genetically altered Bt crops

caging the queen: to confine the queen so that she is unable to lay eggs

candy plug: the sugar candy found in one end of a queen cage; part of the slow-release mechanism of the queen cage

cap: the "lid" made by bees to seal cells full of honey, or that seals brood into the cells to pupate

caste: the distinct types of bees found in a honeybee colony: queen, worker, drone

cleansing flight: a late winter/early spring flight taken by bees who have been clustered in their hive due to cold temperatures during winter—typically notable for the orange-brown spots of bee poop found on snow afterwards

cluster: the tightly gathered ball that bees form in order to survive cold temperatures

Colony Collapse Disorder (CCD): the name given to the phenomenon that began in 2006 where entire hives were found empty. It was thought then this was a specific disease; now it is used to describe the many interconnected forces negatively affecting bees

comb: the wax panels built by bees to raise their young and store their honey constructed of hexagonal cells

comb collapse: comb that has spontaneously detached from the top bar—usually due to heat or poor adherence

coumaphos: the active ingredient in Checkmite+®, a miticide used by beekeepers

cross comb: comb that the bees have not built along the comb guide of the hive's top bars; combs constructed in such a way that they connect multiple top bar bars together making it impossible to inspect the hive

cull: to remove combs that have reached an age that they may contain an accumulation of pesticides

dancing: a method of communicating among honeybees by dancing a certain pattern said to describe the distance and direction from the hive to food sources

dearth: a lack or scarcity of something; in beekeeping, a period of time during which there is no nectar flow

dinks: informal term for a small, less than vigorous hive

drawn comb: beeswax comb built to the full size of the hive's interior; having hexagonal cells of a depth appropriate for the comb's purpose

drawn out: a description of comb where construction of which has progressed to the point where they are sheets built of full depth hexagonal cells

drone: the male honeybee

dry capping: honey that has been capped with a tiny airspace behind the cap, giving the honey a dry white appearance (see cap)

festoon: to hang in a catenary curve, part of the process of bees making natural wax; see *bridge*

fluvalinate: the active ingredient in Apistan®, a miticide used by beekeepers; a synthetic pyrethroid, fat-soluble so it accumulates in beeswax comb

fondant: solid sugar product used as supplemental feed during winter. Useful because it can be suspended from a top bar and placed next to the cluster.

forage: the food that bees fly in search of; the act of searching for said food

foundation: artificial sheets of beeswax or wax-coated plastic intended to assist honeybees in the making of their beeswax comb

genetically engineered (GE): an organism whose genetics have forcibly incorporated the genes of different organisms that are not sexually compatible, using recombinant DNA techniques

genetically modified organism (GMO): an organism whose genetics have been purposely guided over time using natural sexual reproductive processes

honeycomb: wax combs which contain the honey stores in a beehive

honey gate: a valve installed in a honey harvest kit to assist with bottling honey

hot bees: defensive bees; having a tendency to sting with little provocation

integrated pest management: a common sense, environmentally sensitive approach to managing pests; with a focus on prevention of the pest problem, viewing the use of pesticides as a last resort

interchangeable hive equipment: hive parts that fit in all other hives of its type; crucial when beekeepers work together to expand natural beekeeping by making nucs and splits.

lipophilic: tending to combine with or dissolve in lipids or fats

liquid honey: honey that has been extracted from the comb to be used as liquid

marker: a symptom; something that shows the presence or existence of something

masticate: chew; what the bees do to secreted wax to create comb

miticides: a pesticide designed to act specifically on mites

monoculture: cultivation of only a single crop in a given geographical area

Nature Deficit Disorder: phrase coined by Richard Louv in his 2005 book *Last Child in the Woods* describing the effect of human beings, especially children, spending less time outdoors resulting in a wide range of behavioral problems

neonicotinoid: a systemic, nicotine-based, agricultural insecticide

nuc: short for nucleus, a name for a tiny starter hive containing enough comb, bees, brood, food, and a queen to start up a new colony

nuc box: a box of the appropriate size to house a nucleus colony

opening the brood nest: adding blank top bars between existing combs in an effort to slow the swarm impulse by lessening the bees' sense that they are crowded in the hive

orientation flights: brief, hovering flights by young bees as they orient to the location of their hive

overspring: to survive not only winter, but also the early spring

overwinter: to survive the winter

oxalic acid (OA): an organic acid found in many plants, also found in honey; it is not lipophilic, hence it does not build up in the bees' wax comb; recently registered as a miticide

pay it forward: a term used to describe an act that helps another in pursuit of a desired cause. In this case, helping to launch another natural top bar beekeeper by being able to provide them healthy bees as an outcome from your own healthy bees.

phoretic mites: mites found outside the brood cells in a hive; attached to and transported by bees

preempting: to preempt a swarm is to split the hive before they have flown in a swarm

queen cage: the small wooden cage, screened on one side in which a queen is transported either separately, or inside a bee package. Typically contains a candy plug blocking its exit, which acts as a slow release mechanism when starting a new hive

queen catcher: a small slotted container fitted with springs that can be opened to contain a queen bee; the slots are sized so that worker bees can exit but the queen bee cannot

queen cells: the peanut-shaped vertical wax cell where queen bees are raised

queen cup: the early stages of a queen cell; sometimes left empty or torn down without being used

queen-less/ness: a hive whose queen is dead

remainder hive: the colony that remains behind in the original hive when a colony has swarmed

render: to melt and strain wax combs, making clean beeswax which can be used for candles and other purposes

re-queen: destroying the existing queen and replacing her with a queen of different genetics; often used to manipulate a hive for specific genetic traits or to calm a hot hive

ripe honey: honey that has been stored in each cell; capped by the bees and considered ripe when the moisture level has reached approximately 18%

robbing: when a strong hive of bees fights to overcome a weaker hive in order to steal their honey; generally seen during a dearth or drought situation

side center entrance hive: a top bar hive with its entrances in the center of the long side of the hive

slow release: a method of releasing a queen into a hive that involves having a candy plug in the entrance of the queen cage; this slows her release and gives the hive time to acclimate to and accept the new queen's pheromone

splitting: intentionally dividing a hive of bees either to prevent swarming or to purposely increase the size of the beekeeper's apiary

starter kit: a piece of brood comb attached to the anchor bar of the hive; intended to prevent the bees from absconding

survivor stock: bees that have survived a winter in the beekeeper's locale

swarm: the reproductive process of honeybees. Usually presents as a large, noisy number of bees gathered together in a clump, before making their final move into a new cavity of their choosing

swarm trap: any container designed and placed to attract a swarm of honeybees to move into it

synergistic effects: the combined effects of pesticides which are often greater than the sum of the individual pesticides; may create by-products even more toxic than the individual pesticides themselves

systemic pesticides: pesticides that are absorbed by a plant when applied to seeds, soil, or leaves; rendering the entire plant toxic to the insects that feed on them

terroir: that particular set of characteristics such as taste, color and smell that is specific to the locale where it was made

unidirectional: managing a hive so that the bees build their comb in one direction only, moving from brood nest to honey stores

varroa mite destructor: a external parasite found in honeybee colonies; varroa mites suck the blood of brood and adult bees, weakening the bees and shortening their life and spreading disease

wax box: a container designated as a collection box for bits of beeswax comb; should have a tight-fitting lid

wax pot: a pot designated for rendering wax by boiling combs in water; generally not useful for other purposes afterward

wet capping: honey that has been capped where the honey contacts the back of the cap, giving the honey a translucent glossy appearance (see cap)

worker: the female honeybee

Index

About the Author

No sooner was *The Thinking Beekeeper* published than the need for a sequel began to make itself apparent. As public awareness of honeybees and beekeeping increased, the focus began to shift from how to get started with top bar hives to how to manage successful top bar hives through their second year and beyond. *Advanced Top Bar Beekeeping* will guide Thinking Beekeepers through managing an overwintered top bar hive, and preparing for successful splits and swarm management in established hives.

Through her company, Gold Star Honeybees (goldstarhoneybees.com, 207-449-1121), Christy is working to set a viable "gold star" standard for top bar hive equipment interchangeability. Interchangeability is critical for the sustainability of top bar hives—and creating a network of beekeepers using interchangeable equipment will soon allow top bar beekeepers to pay it forward by sharing splits with other beekeepers to naturally launch new hives.

To support top bar hive beekeepers who are interested in connecting with other like-minded folks using top bar hives and supporting the bees' natural systems, Gold Star Honeybees now sponsors Facebook groups designed to help top bar hive beekeepers locate each other. Join the global group at facebook.com/groups/TopBarHiveBeekeepers, and we can then connect you to a more local group for each US state and for several countries as well.

Christy celebrating among top bar hives in northern Italy.
Credit: Paolo Fontana.

A Note About the Publisher

NEW SOCIETY PUBLISHERS (**www.newsociety.com**), is an activist, solutions-oriented publisher focused on publishing books for a world of change. Our books offer tips, tools, and insights from leading experts in sustainable building, homesteading, climate change, environment, conscientious commerce, renewable energy, and more — positive solutions for troubled times.

Sustainable Practices for Strong, Resilient Communities

We print all of our books and catalogues on **100% post-consumer recycled paper**, processed chlorine-free, and printed with vegetable-based, low-VOC inks. These practices are measured through an Environmental Benefits statement (see below). We are committed to printing all of our books and catalogues in North America, not overseas. We also work to reduce our carbon footprint, and purchase carbon offsets based on an annual audit to ensure carbon neutrality.

Employee Trust and a Certified B Corp

In addition to an innovative employee shareholder agreement, we have also achieved B Corporation certification. We care deeply about *what* we publish — our overall list continues to be widely admired and respected for its timeliness and quality — but also about *how* we do business.

For further information, or to browse our full list of books and purchase securely, visit our website at: **www.newsociety.com**

New Society Publishers

ENVIRONMENTAL BENEFITS STATEMENT

For every 5,000 books printed, New Society saves the following resources:[1]

23	Trees
2,041	Pounds of Solid Waste
2,245	Gallons of Water
2,929	Kilowatt Hours of Electricity
3,710	Pounds of Greenhouse Gases
16	Pounds of HAPs, VOCs, and AOX Combined
6	Cubic Yards of Landfill Space

[1]Environmental benefits are calculated based on research done by the Environmental Defense Fund and other members of the Paper Task Force who study the environmental impacts of the paper industry.

IT Solutions Series

E-Commerce Security
Advice from Experts

Mehdi Khosrow-Pour, D.B.A.
Information Resources
Management Association, USA

CyberTech Publishing
Hershey • London • Melbourne • Singapore

Senior Managing Editor: Jan Travers
Managing Editor: Amanda Appicello
Development Editor: Michele Rossi
Copy Editor: Maria Boyer
Typesetter: Jennifer Wetzel
Cover Design: Lisa Tosheff
Printed at: Yurchak Printing Inc.

Published in the United States of America by
CyberTech Publishing (an imprint of Idea Group Inc.)
701 E. Chocolate Avenue, Suite 200
Hershey, PA 17033 USA
Tel: 717-533-8845
Fax: 717-533-8661
E-mail: cust@idea-group.com
Web site: http://www.idea-group.com

and in the United Kingdom by
CyberTech Publishing (an imprint of Idea Group Inc.)
3 Henrietta Street
Covent Garden
London WC2E 8LU
Tel: 44 20 7240 0856
Fax: 44 20 7379 3313
Web site: http://www.eurospan.co.uk

Library of Congress Cataloging-in-Publication Data

E-commerce security : advice from experts / Mehdi Khosrow-Pour, editor.
p. cm.
Includes bibliographical references and index.
ISBN 1-59140-241-7 (softcover) -- ISBN 1-59140-242-5 (ebook)
1. Electronic commerce--Security measures. I. Khosrowpour, Mehdi, 1951-
HF5548.32.E1864 2004
658.4'78--dc22

2003025969

British Cataloguing in Publication Data
A Cataloguing in Publication record for this book is available from the British Library.

All work contributed to this book is new, previously-unpublished material. The views expressed in this book are those of the authors, but not necessarily of the publisher.

IT Solutions Series: E-Commerce Security

Advice from Experts

SECTION I INTRODUCTION

SECTION II EXPERT OPINIONS

SECTION III EXPERT WRITINGS

SECTION IV GLOSSARY AND INDEX

Foreword

From a distance, the concept of e-commerce security seems simple. Just allow authorized people to transact business securely and efficiently through the Internet, and keep unauthorized people away from valuable information. But in today's impersonal and global economy, how can a business or organization really know who they are really allowing into their systems? And how can they be sure unauthorized people are always kept out?

In a highly interconnected and transaction-driven world, deciding who should be kept out or included is becoming more difficult every day. Due in part to interdependent global economic conditions, international terrorism concerns and human ingenuity involved with misusing technology for ill gotten gains, e-commerce security is neither simple nor static.

The managers and executives of companies and organizations that offer e-commerce access points must realize that security is no longer a part-time activity, performed when the IT staff has time each week to check a few system access logs and monitor unused network firewall ports. Around-the-clock monitoring using a combination of automated and human resources has become not only a good marketing story, but is now a business requirement to minimize financial losses and litigation. Inherently this means a dramatically different perspective from a few years ago about what the cost of, and mission focus should be, for the e-commerce security organization.

The definition of what "e-commerce" really means to the business or organization, its clients and its supplier/partners must also be reviewed for scope, clarity and security access. In the rush to be first-to-market a few years ago, important security features were often left out to make the schedule, and have not been addressed since. In other cases, the e-commerce front-end to back-office databases remain open to unauthorized access due to incomplete security architectures, dependence on computer operating system manufacturers, and/or perceived client "ease of use" features and functions.

Compounding the organization's e-commerce business processes and IT budgets is the dearth of industry security standards and plethora of often incompatible proprietary software and hardware products. Continually guessing at which technologies (and their developers) will survive the current market downturn has made selecting the best products suitable for their use very difficult. And last, while e-commerce security is required to protect business assets, as a general rule it is an overhead cost, not a revenue generator. As companies seek to reduce costs, security is often an area of downsizing due to a perceived risk of acceptable financial loss.

As mentioned earlier, e-commerce security is neither simple nor static. Elements of the complex answer are contained in the details of system management, business processes and security technology. Taking the unique approach of interviewing industry experts working on, and solving, complex e-commerce security problems and assignments, this book provides illuminating ideas and discussions about what management and industry practitioners can do to protect their companies and organizations based on best practices and what works. By applying these ideas to specific situations, management can greatly reduce the cost and time required to harden their systems to unapproved access and undesired financial risk.

The U.S. Government reports that several billions of dollars are lost every year to e-commerce and computer security crimes (2002 Computer Crime and Security Survey Report)–an amount growing every day. Stopping that growth rate and driving it to zero will become a larger and larger management challenge and responsibility from the legal and financial perspectives.

Most books that have been written about e-commerce security lack a key ingredient–actual business examples from industry practitioners. This book seamlessly incorporates specific expert information throughout various chapters and sections, providing the reader with relevance and rare insights into current security threats and defenses faced by business leaders.

Compared to an industry research monologue, readers benefit from the experts' multi-decade experience dealing with e-commerce problems: security threat priorities, personal privacy laws, risk management and other at-risk tasks that impact revenue generation and collection. Usually obtainable only at extraordinary expense, readers can learn from and leverage industry best practices, thus saving significant time and expense, while reducing security risks for their organizations.

In the end, e-commerce security is not only about buying and installing technology and stopping unauthorized user access. It is about creating continuous business processes that wrap around and balance end-to-end system and user access, provide transaction security, and require proprietary and customer information to be considered valuable business assets.

Lawrence Oliva
Director
Infrastructure Engineering Program Management
CSC PRIME Alliance, USA

Preface

Organizations of all types and sizes around the world rely heavily on technologies of electronic commerce (e-commerce) for conducting their day-to-day business transactions. Obviously these technologies and all information assets that they process should be protected against fraud, theft and misuse both from internal and external threats. Providing organizations with a secure e-commerce environment is a major issue and challenging one in today's Digital Economy. Without total secure e-commerce, it is almost impossible to take advantage of the opportunities offered by e-commerce technologies. Furthermore, without secure e-commerce applications and practices, it is very difficult to gain the confidence and trust of the consumers and clients in using this technology. Security architecture must be designed to protect organization's e-commerce operations from both known international and external threat, and must be flexible enough to stop less-defined and projected threats. One important element of e-commerce security is top management understanding of the significant importance of the issue and of the commitment given to develop and operate a totally secure e-commerce environment.

E-Commerce Security: Advice from Experts covers a wide range of existing e-commerce security issues and challenges, and offers many solutions for managing security risks. Experts in the field of e-commerce and e-commerce security offer insight to key issues of current e-commerce security and future challenges facing e-commerce professionals in maintaining a secure e-commerce environment. The following paragraph describes the essence of the chapters in this book.

Chapter Summaries

Chapter I, *An Overview*, by Mehdi Khosrow-Pour of the Information Resources Management Association, is an introduction to the topic of e-commerce security. Khosrow-Pour discusses the benefits of e-commerce including online transactions and the sharing of valuable information. The chapter also discusses the risks of e-commerce and the need for a security infrastructure to prevent damages caused by criminal activity. By using a Total E-Commerce Security Program, businesses can protect themselves from dilemmas such as the theft of essential client and business information. Khosrow-Pour outlines what is entailed in building and maintaining a secure e-commerce environment.

Chapter II, *Learning from Practice*, consists of interviews conducted with eight leading e-commerce security experts around the world. In this chapter, they share their experiences and knowledge in the field of e-commerce security. Topics discussed include secure payments, interoperability, consumer/client protection and future challenges facing e-commerce professionals.

In Chapter III, *How One Niche Player in the Internet Security Field Fulfills an Important Role*, Troy Strader, Daniel Norris and Philip Houle, all from Drake Univeristy, and Charles Shrader from Iowa State University, examine the efforts of Palisade Systems to improve the potential of their products in the changing e-commerce environment. Issues faced by Palisade Systems include taking advantage of recent legislation passed regarding privacy on the Internet, the growing need for security on the Internet, and the growth of security products on the Internet.

Chapter IV, *Personal Information Privacy and EC: A Security Conundrum?* by Edward Szewczak of Canisius College in Buffalo, covers the issues of personal information privacy (PIP) and e-commerce. The main focus of this chapter is privacy and barriers in the enforcement of privacy protection. Szewczak's chapter discusses how Internet users are monitored without their consent and how this information could possibly be misused.

Chapter V, by Michelle Fong of Victoria University, is titled, *Developing Secure E-Commerce in China.* China is known for using e-commerce as a growth and modernization in their country. This chapter looks at why business-to-consumer (B2C) online transactions are low. For example, the chapter shows that there are inconvenient and insecure forms of electronic payment. In order for e-commerce to grow, security must improve and earn the trust of the user. Fong's chapter discusses solutions and recommendations for this problem.

Chapter VI, *Identifying and Managing New Forms of Commerce Risk and Security*, by Dieter Fink of the Edith Cowan University in Australia, focuses on Risk and Security Management (RSM). Fink discusses the risks of unprotected e-commerce systems and what solutions offer protection from these risks. Some solutions include firewalls, digital signatures and encryption. The chapter covers new forms of risk and security that are dangers to a system. The chapter shows how to make an effective RMS approach to an e-commerce system.

Chapter VII, *E-Commerce Security and the Law*, by Assafa Endeshaw of the Nanyang Business School in Singapore, emphasizes the need for a secure network. This chapter explains why it is essential to have clear, easy-to-implement regulations and policies. In order to provide a secure e-commerce environment and gain trust of users, organizations and legal departments must get involved. The chapter discusses various laws that have been created to protect a user from identity theft, how offices have been created to investigate and prevent hacking, and talks about the crackdown of unauthorized use of intellectual property. The chapter concludes with solutions and recommendations to the problems discussed.

Chapter VIII, *Rethinking E-Commerce Security in the Digital Economy: A Pragmatic and Strategic Perspective*, by Mahesh Raisinghani of the University of Dallas, discusses the importance of management understanding of advanced technology. Knowledge of advanced technological systems assists in creating more robust, scalable and adaptable information systems for the organization dedicated to continuous improvement and innovation. The key questions covered in this chapter include assessing status quo and modus operandi. Raisinghani proposes alternative solutions and recommendations for the future of the digital economy.

Chapter IX, *Security and the Importance of Trust in the Australian Automotive Industry*, by Pauline Ratnasingam of Central Missouri State University, focuses on the Australian Automotive Industry and security issues in technical, political and behavioral perspectives. The chapter discusses the importance of trust, how to gain trust and how to establish dependable business relationships as a means to mitigate risks in EDI in the Australian automotive industry.

Chapter X, by Daniel Ruggles from Consulting Associates, LLC, is titled *E-Commerce Security Planning*. In this chapter, Ruggles discusses the importance of establishing an e-commerce trust infrastructure that will protect three major areas of an e-commerce site: the internal network, the perimeter network access and the external network. The chapter discusses how to build an e-commerce trust infrastructure and how to create a balance between functionality and accessibility with an e-commerce system.

As emerging e-commerce technologies offer new opportunities to organizations all over the world to conduct business in ways that until several years ago was considered unthinkable, it is extremely important to protect both the organization and its clients from fraudulent acts and misuse of these technologies and information that they process. Without a totally secure e-commerce environment, all efforts to obtain the trust and confidence of the user communities will be thrown to the waste and will not bear any meaningful results. Planning and conducting totally secure e-commerce should be the first step in developing any kind of e-commerce applications. For those organizations that have already been involved in using e-commerce security, they should continuously reassess their e-commerce security and to make sure that all vulnerabilities are planned for and possible threats are also planned for. Achieving a totally secure e-commerce environment begins with a well-informed management about the significance of secure e-commerce. I hope that the knowledge and advice provided in this book can become instrumental in understanding many issues, including those of e-commerce security, and can provide insights and ideas that can be utilized in support of e-commerce security. As always, your valuable comments and feedback will be greatly appreciated.

Mehdi Khosrow-Pour, D.B.A.
Executive Director and President
Information Resources Management Association, USA

Acknowledgement

This book is a result of cooperation between expert academics and corporate professionals specializing in e-commerce security. I am very grateful to all those who accepted my invitation to participate in this important project. Without their knowledge and expertise, this book could not have been published. I would also like to take this opportunity and express my many thanks to my assistant, Jennifer Wetzel, for all her hard work and for keeping me and the book project on schedule. Further thanks are due to the CyberTech Publishing editorial team, Jan Travers, senior managing editor and Mandy Appicello, managing editor. Finally, this book is dedicated to my two little girls, Bosha and Anar.

Section I

Introduction

Chapter I

An Overview

Mehdi Khosrow-Pour, D.B.A.
Executive Director and President
Information Resources Management Association, USA

Biography

Dr. Mehdi Khosrow-Pour is currently the Executive Director and President of the Information Resources Management Association (IRMA) and Senior Technology Editor for Idea Group, Inc. He is also the Editor in Chief of the Information Resources Management Journal *(IRMJ),* The Journal of Electronic Commerce in Organizations *(JECO), the* Annals of Cases on Information Technology *(ACIT),* Information Management *(IM), and Consulting Editor of the* Information Technology Newsletter *(ITN). He is the author/editor of more than 20 books on various topics of information technology utilization and management in organizations, and more than 50 journal articles.*

Introduction

During the past two decades, the business world has witnessed a technological revolution known today as electronic commerce or e-commerce. This revolution has allowed businesses all over the world to conduct business in ways that were unimaginable two decades ago. Through the use of e-commerce technologies, businesses can share and disseminate information electronically and conduct business online so consumers, regardless of their locations, can obtain goods and services from the businesses. Because of the many opportunities e-commerce technologies offer in today's competitive marketplace, it is essential for organizations to have e-commerce presence and effectively utilize the Internet to expand their businesses. With this Internet presence, ensuring security of their data and sales experiences is of paramount importance. Through the use of effective e-commerce security tools, business can increase their sales, reduce the cost of doing business, and at the same time increase customer service and satisfaction. These objectives cannot be easily achieved if the organization does not have a clear business strategy for its e-commerce initiatives and operations

E-Commerce Security

Although e-commerce technologies offer immense benefits, conducting any kind of online communications or transactions offers the potential for greater misuse of these technologies and even potential criminal activities. The issues of technology security and misuse are not only limited to e-commerce technologies, but rather are part of much broader issues affecting computer and information systems throughout the world. Each year, many organizations become the target of security-related crimes ranging from virus attacks to business fraud, including the theft of sensitive business information and confidential credit card information. These security attacks cost businesses millions of dollars and interrupt their operations. Many reporters and consultants estimate the cost of damages related to security breaches as billions of dollars. However, it is not the issue of the accuracy of the range of damages, but rather the fact that with the ever-increasing number of users of information systems, easy access to information, and the increasing number of knowledgeable users, one can easily assume that the number of technology misuses and security threats will increase proportionally.

Regrettably, the true extent of damages incurred by businesses related to information security crime cannot truly be known due to the fact that many businesses are reluctant to admit that their systems were infiltrated, and to share knowledge about the incident and the extent of the damage. The reluctance of businesses to share information regarding their security breaches is based on a common fear that by allowing the public to learn about the incident, their customers will lose confidence in the business's ability to protect its assets and the business will lose customers in turn losing their profitability. Because today's consumer is leery of providing financial information online, the business has nothing to gain by voluntarily admitting to having been victimized by security-related crimes. In today's media frenzy regarding the Internet and its functions, maintaining a positive image regarding e-commerce security is the number one concern of many businesses and completely necessary to maintaining the business's survivability and competitive stance.

The lack of the first-hand knowledge of the actual cases makes it much more difficult to plan and deal with information security threats. However, during the past decade, information security technologies and methodologies have improved considerably, as have overall management practices in planning and protecting organizational information technology resources. There are now experts specializing in the field of cybersecurity, which has much to do with all activities related to the protection of e-commerce technologies and practices from potential criminals in cyberspace. In addition to security measures designed to protect information technology resources, many businesses are realizing that being successful in e-commerce endeavors requires significant investment and planning in developing a total security program to protect their Internet investment and to prevent criminals from infiltrating their systems and causing damage to the e-commerce endeavors.

Total E-Commerce Security Program

A total e-commerce security program consists of protection programs that utilize available technologies (hardware and software), people, strategic planning, and management programs designed and implemented to protect the firm's e-commerce resources and operations. Such a program is essential to the overall survivability and operations of the business's e-commerce efforts, and the organization should consider it an integral component of successful e-commerce strategy. Success of such a program depends on the complete support of top management and full participation of both the IT department and management in understanding the effectiveness and limitations of the program. In addition, such a program requires continuous assessment and evaluation to make sure that the program and its tools and solutions are up-to-date and have incorporated the latest technologies and management practices.

Assessing E-Commerce Operations

The first step toward developing a total e-commerce program is to conduct a full assessment of the value and importance of e-commerce in the overall success of the firm's business plan and objectives. The next step should be to assess the vulnerability of the firm's e-commerce system both in terms of threat posed to the system internally and the risks that exist externally. External vulnerabilities are much easier to identify, and internal vulnerabilities are more difficult to see. However, all attempts should be made to identify any areas in which the system can be misused internally. This task can be greatly accomplished by talking to both system users and developers. In addition, there are many software packages out there that will monitor the overall use and access of the system internally, and provide full analysis and possible suspicious activities of the internal users.

Continuity Plan

The next step is to begin developing an e-commerce continuity plan that will clearly map out all possible shortcomings, ways to prevent and deal with them, and contingency plans for recovering from security threats and breeches. Many businesses tend to believe that having anti-virus programs on their systems, and firewalls installed, totally protects their systems. These are excellent beginning steps; however, even with these measures, e-commerce systems still face the following shortcomings: (1) fire/explosion; (2) intentional destruction of hardware, software, and data/information; (3) theft of hardware, software; (4) loss of key e-commerce security personnel; (5) loss of utilities; (6) loss of technology; (7) loss of communications; and (8) loss of vendors. Threats to each one of these areas should be carefully assessed, and contingency plans should be developed which explain in detail how to deal with each potential shortcoming. Furthermore, the plan should identify individuals responsible in taking the lead in correcting the problems associated with the shortcomings.

E-Commerce Technologies

Next, organizations should assess the available hardware and software protecting the firm's e-commerce system. It is important to understand that the technologies used must be adequate to meet the needs of the firm's system and provide the level of protection for all of the possible security threats. Critical areas to consider include: the sensitivity of the data/information accessed, the volume of access traffic and methods of access. Technology-based e-commerce security should consist of different layers of security by utilizing either SSL (Secure Socket Layer) or SET (Secure Electronic Transaction) security schema to provide full security in online transactions. The overall goal of technology-based security should be to provide effective authentication, integrity, encryption and non-repudiation. During this stage of developing the total e-commerce security program, many firms utilize the expertise of outside firms that are specialized in assessing technology-based security systems.

People

The most important component of any effective security program is the people who administer and operate the program. Security breaches are not committed by systems but rather by people who mange them and their users. Most studies in the past have indicated that the internal threats to e-commerce systems are often much greater than the outside threats. In many cases, the criminals who succeed in breaking into a system have either prior knowledge of the system or have internal co-conspirator. The most important tool in the hand of management in reducing the internal threats is educating all parties involved–both the internal user community as well as system management personnel–regarding the consequence of violating the system security and integrity. Many users are aware of the fact that breaking into an information system is considered a Federal crime, and violators can be prosecuted by law. By informing users, the firm creates a certain degree of deterrence that will discourage many not to violate the law.

Strategy

Developing a sound and effective strategic plan for e-commerce security is an important task. Such a strategy should include: the overall goal of the total e-commerce security program, its objectives and scope. This strategy should be in line with the overall e-commerce business strategy of the firm. One of the objectives of this strategy should be to provide full protection for all e-commerce resources of the firm and to allow the business to recover from any incident of operation interruptions as quickly as possible, minimizing the revenue losses and cost of recovery. The strategic plan should also include the needed resources in support of the goals and objectives, as well as the constraints and limitations facing this strategic plan. Additionally, the strategic plan should also include key human resources, management structures and decision-makers in implementing various e-commerce security programs developed as the part of the strategic plan.

Management

The success of a total e-commerce security program depends on the effective management of such a program. The management support begins from top management and continues to all other levels of management. Such a program must be directly managed and monitored by a senior manager in the case of a larger organization and perhaps directly managed by the owner or president of medium or smaller companies. The primary responsibility of the program manager should be to keep the strategic plan fully up-to-date, implement its programs and continuously reassess the effectiveness of the existing programs. Part of this person's responsibility should be to learn from the effective practices of other organizations regarding their e-commerce security programs. The knowledge can be gained through reading published case studies, books on the subject, as well as published articles.

Conclusion

Lack of an effective e-commerce security program can be disastrous and costly, leading to loss of business revenue and adversely impacting the integrity of the firm's e-commerce operations and reputation. It is crucial for all businesses utilizing e-commerce technologies to devise and implement effective security measures that will provide the highest level of security to their e-commerce resources. In many cases, the costs of preventing security problems are much less expensive than the costs of recovering after the firm has been victimized. This requires careful assessment and planning of all critical factors related the firm's e-commerce security. Regrettably, in the midst of maintaining the existing e-commerce security system, sometimes managers overlook the need to improve important aspects of security programs or they mistakenly believe that "it won't happen to us." The manager's oversight can allow the firm's system to become vulnerable to misuse and attacks.

Recommended Readings

Lim, E. (2003). *Advances in Mobile Commerce Technologies.* Hershey, PA: Idea Group Publishing.

Liu, C., Marchewka, J. & Mackie, B. (2003). Implementing Privacy Dimensions within an Electronic Storefront. In J.R. Mariga (Ed.), *Managing E-Commerce and Mobile Computing Technologies* (pp. 116-131). Hershey, PA: IRM Press.

Sharma, S.K. & Gupta, J.N.D. (2003). Adverse effects of e-commerce. In S. Lubbe & J.M. van Heerden (Eds.), *The Economic and Social Impacts of E-Commerce* (pp. 33-49). Hershey, PA: Idea Group Publishing.

Singh, M. (2004). *E-Business Innovation and Change Management.* Hershey, PA: Idea Group Publishing.

Thanasankit, T. (2003). *E-Commerce and Cultural Values.* Hershey, PA: Idea Group Publishing.

Section II

Expert Opinions

Chapter II

Learning from Practice

Introduction

To learn more about the views and practices of industry practitioners, this chapter outlines the result of an interview with a panel of e-commerce security practitioners who were asked to share their insights, understanding and vision regarding issues related to the practice of e-commerce security. Panel members were chosen based on their experience in the field of e-commerce security and management.

Section I of this interview deals with the issues and challenges of e-commerce security. Issues covered in this section include e-commerce security policies, tactics of hackers and security architecture. In Section II of the interview, the participants were asked to provide suggestions and recommendations regarding current challenges, solutions and future issues facing e-commerce security. Panel participants were asked to answer each of the questions to the best of their knowledge, sharing their practical experiences and understanding regarding e-commerce security with the book's audience.

For the list of panel members and their profiles, see Appendix A at the end of this chapter (pg. 67).

Section I: Issues and Challenges

Q: Does your organization have a formal e-commerce security policy in place?

Naglost:

Being a provider of e-commerce and Internet services to both government and private sector organizations, Berkeley must have a formal e-commerce security policy, as we often create and implement these solutions.

Oliva:

We also have one in place. Our security policy uses established software technologies (firewalls, encryptions, etc.) and formal business practices such as "know the source" and shredding of unneeded documents.

Thompson:

We have an e-commerce security policy in place, as well.

Arazi:

There is one in place.

Updadhyaya:

We do not have a policy and there is no plan to develop one. Our company does not have the resources to set up and maintain a formal e-commerce security policy. It is something that we want and would love to have, however, it is not feasible at this time. When should others or we adopt one? I think as a company you need to be realistic about when this is implemented. Once your e-commerce venture has started to produce revenue, you can start to spend the resources on developing a formal plan. However, don't be reactive, start making small chunks of plan, target certain areas as you need them. Then combine them, refactor them and use that as the basis of your formal policy once the company is ready.

Thomson:

There is one in place at Beanstream.

Mahmood:

Yes there is one in place at the University of Texas, El Paso.

Seen:

There is currently one under development at Murchison. We found that developing a policy can be a relatively quick procedure–it is the implementation of the policy which requires a commitment of time and effort, and of course, buy in from management and those who control the purse strings of the organization. The benefit of having an ingrained security policy is that it helps focus our attention on one of the key issues in the industry, maintaining and increasing the level of trust people have in e-commerce. This allows us to directly assist our clients in bringing their e-commerce solutions online while still being able to sleep at night.

Q: Is your organization's top leadership supportive of having an e-commerce security policy? And in your opinion, how can e-commerce managers educate and gain support from top management regarding this important issue?

Naglost:

Our top leadership is certainly supportive of the e-commerce security policy–we are willing to set aside both time and resources to ensure the policy is adequate in terms of breadth and depth, and is implemented accurately and completely throughout the organization.

Arazi:

Our management is very supportive, understanding the need to have clear policies and procedures on how information is collected and processed, as well as how it is accessed and protected. I believe that e-commerce managers need to employ a dual-aspect process of educating top management about the benefits of having such a policy in place versus the risks of not.

Thompson:

Our organization's top leadership is supportive, too, although, as it is outside of the day-to-day operations of a university, it is up to the management and team members of CECC to develop and explain policies to management to educate them and gain support for important issues and investments in this area.

Oliva:

I would say our top management supports e-commerce security activities as well, though more from a defensive and offensive perspective. E-commerce managers must use accurate business data to show management the financial consequences (revenue losses, recovery costs, litigation costs) of not implementing information and e-commerce security business practices.

Thomson:

We support e-commerce security at all levels in our organization. In my opinion, the easiest way to convince management that they need to address the issue is to simply provide examples of other organizations that have not, and the negative publicity and customer support issues that are generated as a result.

Upadhyaya:

Our leaders fully understand the need to be secure, but we are also fortunate that the MD of the company was UK's digital content forum chair (DCF) for 2002. The biggest motivation for them is liability. As soon as it was explained that taking financial payments over the Internet (even B2C) was not without risk, their ears pricked up. In terms of cost, it is much more efficient to be prepared rather than trying to recover from an attack. Sadly, top-level management see things in the terms of cost, so explaining the potential damage is a sure fast way of getting support.*

** The DCF was created to develop the digital content sector by forming a two-way conduit between industry and government to gather views and input into policy-making practices (www.dcf.org.uk).*

Mahmood:

Our top management is absolutely aware of the need for an e-commerce security policy, but I think it's important to add that the e-commerce manager should give seminars on the issue. The manager should keep the top management aware of the latest developments and should explain to them why it is important for the company to use e-commerce tools and technologies, and why it is important for the organization to have an e-commerce security policy.

Seen:

The development of our e-commerce security plan has full support from the CEO down; however, in our situation, it was not necessary to "sell" the benefits of security management–we needed to ensure that the policy would be implemented. This means, for example, that a programmer or administrator could immediately drop what they were working on to attend to a security issue even if it is just the application of a critical patch for a system. It took some resistance from the programmers and administrators to ensure we were allowed to respond this way.

Q: In your opinion, how does your organization deal with the tactics of hackers (e.g., altering files, eavesdropping on transactions, sending viruses and cookies, etc.)?

Naglost:

Berkeley implements a multi-tier system of security to deal with hackers. We implement robust firewalls using the latest technology (both software and hardware) available and ensure all operating systems and server and client applications are up-to-date and have all the required security patches installed. We also implement anti-virus software on all servers and workstations and ensure these are kept up-to-date. Our regular monitoring includes checking server logs, utilizing state-of-the-art server monitoring software to ensure we are alerted to any suspicious activity, and subscriptions to relevant security bulletins to ensure that we are aware of the latest known e-commerce security threats.

Oliva:

For general information protection we use commercially available software (firewalls, file scans, etc.) to protect desktop PCs and servers from hackers, denial-of-service attacks and zombie attacks. We advised staff not to download files or open e-mails unless they are sure of who sent them (which is the "know the source" policy). Files are detected immediately if coming from unknown sources.

Thompson:

In our case, those servers that contain client information and applications that are managed by the CECC are housed with a specialist Web-hosting company. The primary responsibility for dealing with the tactics of hackers is therefore passed to this organization.

Arazi:

A2i deals with such tactics by segmenting the IT infrastructure into multiple tiers and having the bare minimum exposed to the Internet. A very strong and closely guarded perimeter network is connected via firewalls to dedicated DMZ networks for the Web servers and application servers, with the database servers further protected in yet another network segment. Sensitive data is encrypted immediately after being processed, and all servers contain only the required applications and have been "locked down" by applying strict access privileges and constant security monitoring. Application-specific filters and proxies handle inbound and outbound mail and Web connections to fight spam and viruses. Multiple layers of frequently updated virus software are also used.

Upadhyaya:

Unfortunately we weren't proactive at first. We have been hit a couple of times, but each one has been an education. The first time was the summer of 2001 with a variant of the Code Red worm. When it was detected that we had slowdown on LAN traffic and echoes to our servers were timing out, we realized there was a problem. As our company pretty much only uses Microsoft technologies, we were aware that we were more at risk from OS flaws and worm attacks than some of our other colleagues. It was noticed that our main Web server was defaced, one of our remote transactional database servers was not functioning properly, and our VPNs/WANs were down. All our development servers had been opened up. The clean up was a logistical nightmare! However we succeeded and it changed our approach to worm style attacks. Worms force computer security to be an everyday battle–each day you need to scan the newsgroups for any new patches or updates. Then take the time to deploy them on all relevant machines (though some OS/ products do allow a live update now). With Internet-connected machines, not only do you have to worry about your own systems, you have to worry about other networks on your exchange. When we recovered from the Code Red variant, it still took a while to return to normal service levels, as other infected networks where flooding our ISP's bandwidth. It was an education; it's hard as a small company to devote resources to be purely pre-empt the information militia (the cracker/hacker communities). I think it's the case of "once stung, twice shy," however that's not really the most efficient way to behave. I think the key is getting your servers to an acceptable level of secureness, then improving your response time. To improve our response time, we set up external monitoring of our sites, checking both the dynamic and static versions, thus targeting the nature of the attack IIS or SQL server. As soon as there's a problem, an e-mail is sent to the relevant parties and a message is also sent to their cell phones. As all the IT staff have broadband at home, updates or fixes can be done remotely, minimizing the duration of any DoS attacks. Currently we block all ports that we don't need and use NAT to transform some of the obvious ports such as 1433 (Microsoft SQL server). Then we use IP security/filtering at the machine level to block anything the firewall has missed. After the transversal attack, we stopped using default directory structures and placed all websites on a new drive. On the SQL server side, if you can, try and use the Windows Active directory to manage your SQL security. Basically the rule is: when you set up your infrastructure, no matter which platform you use, don't use any defaults. Because once you do, you're giving a clue to any information intruder. Even the most novice, bored 14-year-old hacker with a downloaded PDF manual could find entry into your system if you accept the defaults. Since these changes, transversal attacks to gain access to servers file system have ended. Since then, we are patched to the teeth. We use third-party software from eEye and keep up-to-date with OS patches. However, none of this is innovative, it is reliant on another company discovering a new tactic and then letting the world know.

Thomson:

We made a decision early in our business development that Internet security was not an area of core expertise, so we outsourced it. We use a third-party security firm to monitor our firewall and servers 24 days, seven days a week to attempt to detect and defeat any attempt by unauthorized third parties to gain access to our systems or services.

Mahmood:

It is very difficult and time consuming to stay ahead of the hackers. You can use a number of commercially available tools to stay ahead. This includes any tool that uses SSL to provide end-to-end secure communications. To prevent hackers from getting into the system, you may want to have a firewall installed right outside your intranet. Biometric devices are being used to provide security to the sensitive materials.

Seen:

That is why our number one goal is prevention, obviously. We take an active role in monitoring suspicious activity within the network as a whole and try to limit the potential for a breach. Containment is secondary to our way of thinking–if a hacker gets in, even if they are contained, the damage has been done to our reputation for security. Containment and limiting our exposure are still important, but it is better to never be placed in such a situation to begin with an e-commerce security threats.

Q: In your opinion, how important is the issue of security architecture and what should be covered in system security architecture?

Naglost:

A comprehensive and detailed security architecture is paramount to ensuring that all aspects of the organization's information platform are secured and monitored. The system security architecture should cover all internal systems that the organization manages as well as any external systems that interact directly with the organization. The architecture should include general objectives as well as a complete listing of security processes, policies and services. The physical and logical structures of the information systems should be covered as well as the associated levels of trust between each system.

Oliva:

I would agree that the issue of information security is critical to the success and growth of our business; however, there are limits on what we can afford to invest in. There is no direct financial return from having good security practices—just overhead and avoided costs. In terms of architecture, we believe there are four primary elements:

1. Wired Systems Security (local and wide area networks, desktops, servers, printers, etc.)

2. Wired Systems Security (cell phones, palm pilots, Wi-Fi networks, Bluetooth, etc.)

3. Business Practice Security (superfluous document destruction, unsolicited telephone inquiries, "know the source," verify requests twice, not disclosing private information in public places, etc.

4. Software Applications Security (e-mail, e-commerce, customer support, sales systems, etc., secured at the access point and user authority level).

Thompson:

Security architecture may be the most important element of any e-commerce environment. Public key infrastructure (PKI) and the issues regarding its implementation within an organization are particularly important.

Arazi:

I believe security is a process, not a product. As such, an organization's security architecture is of paramount importance as it serves as the "roadmap" for navigating the never-ending route toward increased security. Both threats and counter-measures constantly evolve with time and need to be dealt with on multiple levels: those of the individual systems, their connecting network and organizational processes and policies. A good security architecture will cover the specific issues outlined in each level as well as tackle the interfaces between those levels. Additionally, the security architecture must balance.

A balanced security architecture is one that balances cost of implementation with security. Such implementation costs obviously include budget, but usability and flexibility losses or constraints must also be factored in. Users, both internal and external, must not feel hampered by the application of the security policy, and the organization's business goals must always direct security rather than security concerns directing how business is conducted.

Upadhyaya:

Exceptionally, the underlying security of any system will tend to be the most vulnerable, as it doesn't usually get exposed until after an attack. What's needed is:

1. A set of principles for establishing and maintaining features and mechanisms that protect against interruption and loss to packet-switched network elements, the communication services they provide, and the data they contain and carry.

2. *A testing schema that simulates known hacker tactics.*
3. *System-wide authentication methods, that work from page to page (in a Web context).*

Thomson:

The architecture of security systems is largely beyond my area of expertise as we outsource that capability. I deal with architecture at the operating system level. We have outsourced Internet security to a third party, Presinet Systems. We knew that security was a very important aspect of our business and that we were not experts in this regard. Having a third party monitor our systems provides our clients with assurance that our network is properly managed and that any issues related to security are handled by experts in this field.

Seen:

For us, a secure, verifiable architecture is a mandatory element of our systems. If someone can look at a system on paper and spot a security flaw or weakness, then it is a case of going back to the drawing board, taking a good hard look at what you've got and trying to pinpoint where the architecture is lacking. Too often, more time is spent trying to "optimize" an architecture that is fundamentally flawed from a security point of view and should be thrown away. Security should be a by-product of solid design, just as insecurities are by-products of a flawed design.

Mahmood:

The issue of security architecture is very important. It should cover all the security techniques that are available at the present time. It should say how the company's sensitive information can be secured.

Q: How does your organization deal with the issue of consumer/client protection in its e-commerce environment?

Thompson:

We try to provide comprehensive information to the client on all aspects of our e-commerce environment. In our opinion a consumer is in most cases better protected by being well informed. All e-commerce transactions are secured using 128bit SSL encryption. Any transactions are completed using real-time credit card authentication with the financial institution. This reduces the amount of consumer/client information the needs to be stored.

Oliva:

Our e-commerce site is SSL as wellas Verisign secured for extra transaction protection. We have postings to that effect on the site to let customers know our dedication to the security of their information.

Naglost:

Berkeley protects consumers who use its services by implementing both a technical and legal security and privacy framework. We incorporate explicit privacy and security policies and ensure that all consumer data stored by Berkeley is secured at the maximum level. All e-commerce transactions are carried out in a secure environment and, where possible, personal information that is only required by third parties (e.g., credit card details required by transaction gateways) is transferred directly to the third party and is not made accessible to Berkeley.

Arazi:

In addition to the use of SSL and https, we design our systems to encrypt customer information as soon as possible after its receipt, and further use time-limited pseudo-random record ID locators to thwart unauthorized data-mining operations. Whenever possible, servers that store customer-sensitive information are located behind two or more layers of firewall protection, have very restrictive access policies and elaborate logging.

Thomson:

We use a combination of technologies to protect consumer information within our technical infrastructure. We protect data streams by a combination of SSL, VPN and PGP technologies. Sensitive e-mail information is sent using PGP. Access to merchant information is controlled by organization name, user name and password. We provide the ability to limit access to certain functions by IP address and/or GEO-IP mapping.

Seen:

The usual steps are taken to protect the confidentiality of users in our systems. Obvious steps are things like ensuring basic database security measures are in place (this means NO default installs), only collecting data that is absolutely necessary, making sure a credit card number is never stored but rather passed directly to the processing gateway, things like that. The benchmark is often a subjective one; we ask ourselves, "Would we put our details in here?" and if you can't be confident of the security of your own system, why should other users trust it? I cannot stress enough, however, the basic steps are the most important ones to take.

Mahmood:

Yes, the consumer information protection is very, very important. If the consumer does not have confidence in your security system, they will not do business with you. The latest survey on this issue indicated that security is the number one reason as to why a lot of people do not shop online. We use SSL to secure transactions to protect sensitive client information.

Upadhyaya:

Customer/client confidence is fundamentally the most important issue in any e-commerce transaction. Without protecting that, you have no business. In many ways, the transaction security of a site can be compromised such as spoofing or sniffing. In the cases of spoofing and sniffing, our preferred technique is to use data encryption or signed data for the transaction.

Q: How does your organization deal with the issue of protecting consumer/client identity during an electronic agreement?

Naglost:

Berkeley ensures the protection of consumer identity during an electronic agreement by collecting only relevant and required information and then affording these details the same high level of security as its own corporate data. Consumers' identities remain anonymous when there is no requirement to ascertain or record their identities.

Oliva:

At this time, due to a lack of legal clarity about their validity, we have not started to use electronic identity software for legal agreements—we still use signed (paper) documents and do not use e-documents. Electronic identity software enables the user to transmit their legal signature in an encrypted, indisputable file format. The advantages are clear and numerous to businesses competing on a 24x7 global basis. Dramatic decreases in paperwork shipping costs, immediate approval for legal documents regardless of client or attorney location, around-the-clock customer purchase and approval capability, and reduced paperwork are all real advantages of digital signatures. The software tools were built and tested several years ago, and are ready for immediate use. The major issue concerning electronic identity software involves legal acceptance. To date, no significant legal guidelines have been developed that permit corporate or public agency acceptance of a digital signature in lieu of a signed paper document. In short, due to legal and financial risk, no one wants to be the "pioneer" in the use of digital signatures. Should a future court ruling adversely impact digitally signed legal or financial documents, it is highly probable that organizations could face substantial damages. The trigger for full business and government acceptance of digital signatures will be court cases that build a precedence of legal opinion that with proper safeguards, electronic identity software holds the same legal weight as paper documents. Until such legal activities have occurred—hopefully in the next five years—we advise clients to continue to request approvals made on paper documents.

Arazi:

We endeavor to design and implement systems where customers are required to log in once and be validated, after which a unique, non-linear, time-limited, pseudo-random session ID is generated for their session or transaction. A master, highly secure "authentication server" then stores this information. After successful validation and throughout the entire transaction, customer information is then referred to by this unique identifier rather than by information pertaining to their identity, such that the customer/client identity is completely anonymized to most servers and applications handling the transaction, with each system/application only processing the fields of information it has been authorized for.

Thompson:

All e-commerce transactions at CECC are secured using 128-bit SSL encryption. Any transactions are completed using real-time credit card authentication with the financial institution. This reduces the amount of consumer/client information that needs to be stored.

Upadhyaya:

For us, in terms of non-capture of client information during a transaction, when the client fills in a payment form and submits the data, their details are not sent straight away. What actually happens is that a secure link is set up between the client's browser and our merchant. An encryption code is requested and received, which then wraps the transaction details before leaving the client software.

Thomson:

Any information that we store as a result of an electronic agreement is encrypted. The systems storing the information themselves are housed in a high-security data center. We also deploy the technology solutions listed in Q5 during the agreement. The authentication services that we deploy in order to authenticate parties during an agreement do not display any confidential customer information. Our system scores the results of various decisions, and it is only that score that is ever displayed to a merchant or the client.

Mahmood:

My organization protects the consumer/client by using digital certificates to verify identity. Digital certificates clearly identify the person that you are dealing with. They prevent fraud because they will prevent a fraudulent person from assuming someone else's identity and causing all kinds of mischief.

Seen:

Obviously, overall, we need to balance consumers' rights to privacy with the need to ensure that suitable audit trails are maintained in case of accusations of fraud, detection of a compromise, etc. Murchison ensures that the minimum information necessary is collected and that it is secured within a database that does not directly face the Internet during the processing of payments and transactions.

Q: In your opinion, what measures can a business take to make consumers/clients feel at ease regarding the level of security protection of a business's e-commerce system?

Mahmood:

Use digital certificates in conjunction with SSL.

Arazi:

A business must cause customers to feel comfortable and confident about the business, the manner in which transactions are conducted, and what is done with information that is collected from customers. Unfortunately, what the customers feel have very little to do with what technical precautions are taken. I have found that registration with certification authorities such as eTrust and VeriSign make customers feel safer because the names are recognizable. A privacy policy is helpful as well.

Oliva:

As has been noted, businesses can provide "evidence" that permits the customer to check for him or herself that a transaction is secure. This can be done through the software verifiers mentioned (such as the Verisign seal) along with membership in various trade groups that clarify a website complies with the very few industry security standards.

Thomson:

I, too, think that one of the best measures a merchant or service provider can adopt in order to provide consumer confidence in e-commerce systems are industry best practices such as the Visa AIS program, WebTrust, Verified by Visa, etc. The merchant can leverage the brand recognition of these programs and have a very good security and privacy baseline to work from.

Naglost:

Businesses must provide clear, simple and easily accessible information regarding their security policies and procedures to all consumers using their e-commerce systems. They must dem-

onstrate their level of security protection by providing relevant information to consumers and by providing avenues for consumers to validate this information with a third party (security service provider and/or government organization).

Seen:

And certainly not by touting their systems as "unhackable" or otherwise as we have seen some companies do in recent times. I think the biggest comfort factor in online transaction is giving consumers a name—a real life contact person—even if it is only via e-mail. Customers tend to get suspicious when they are expected to divulge sensitive data on a Website that has no tangible links to a real-world entity. If they feel that they have someone to hold accountable, it makes trusting that person to protect their data just a little bit easier.

Upadhyaya:

It is also important to avoid the number of jumps between servers—it makes clients nervous. Your client is going to wonder what information is being sent back and forth. The best way is to try and provide that full solution without using a third party. However, again, make sure that a reputable company has assured your site.

Thompson:

1. Clearly explain each stage of a transaction to the client and how the information they supply will be used.

2. Ensure that technical issues are explained clearly.

3. Provide links to external information on security issues wherever appropriate. For example, providing a logo and link to the Verisign website if that particular certificate authority is included on the organization's SSL certificate. Whenever a client/consumer is asked to provide information, be it personal, postal, credit card or otherwise, the following should be explained: how the information will be used; how long the information will be kept and for what purposes; and will the information be made available to any third parties? At the beginning of a transaction, it should be explained in clear terms how the information being provided is secured in transit and in storage (if applicable).

At the conclusion of a transaction it should be made clear what the client/consumer should expect to receive and what the process for fulfilling the transaction will be. "In the next few minutes you will receive an e-mail receipt to the address you specified. You can keep track of the progress of your order at the following address http://www ... using your username/password."

Q: In your opinion, do you believe that self-regulation regarding consumer/client protection in an e-commerce environment is adequate or is government involvement needed? What should be included in any regulations regarding consumer/client protection?

Oliva:

After 10 years of industry resistance to even minimal regulations, it is clear that government involvement will be required to enforce even token protection levels for all Internet transactions. This is in large part due to a lack of customer mandates for security and incredible bickering over which standards are "best." In terms of regulations, all e-commerce and/or financial transactions should be secured through software encryption and/or biometric identification to deter theft or illegal use. This could be as simple as adding one hundred extra bytes of identification data to each transaction.

Naglost:

In fact, federal and international government organizations must be involved in the development and enforcement of regulations regarding consumer protection. These regulations must include specific details covering the rights and responsibilities of the businesses providing the goods/services, consumers and any other intermediaries who are involved in facilitating the e-commerce transaction. The regulations must be enforceable across national/international borders.

Thompson:

One outcome of additional government regulation regarding consumer/client protection in the e-commerce environment would be the introduction of additional costs in ensuring compliance. Self-regulation is therefore preferable from a business perspective. However, many consumers are not fully informed and, without adequate regulation/safeguards, they may be exposed to unacceptable risks in an e-commerce environment. Positive public perceptions about the degree of security offered through e-commerce continue as an important element in consumer/client uptake. What should be introduced in terms of regulations remains unclear.

Seen:

The big question is, "Which government is going to regulate consumer protection in e-commerce?" We have already seen the battle for jurisdiction on the Internet played out a few times. Consumer protection is important and regulation is a noble idea, but there is no way that it can be achieved globally in a uniform fashion. I believe that while self-regulation may not be adequate, it is better than no regulation at all.

Upadhyaya:

But, self-regulation is only effective if all parties adhere to the resolution. Industry standards bodies and ombudsman regulatory enforcers would help protect consumer interest. As of August 21, 2002, the UK's Department of Trade and Industry (DTI) set out regulations intended to boost consumer confidence. The regulations also contain requirements about any e-mail advertisement and enable a recipient of an unsolicited advertisement to identify it without opening it.

Thomson:

I don't think that self regulation is viable simply because we have seen the wide range of security solutions that vendors and merchants deploy, many of which are unsuitable. Having said that, I do not think that government regulation is the answer either. In Canada and the UK (two markets that we serve), government has set standards for the protection of personal information and consumer privacy, but they are unevenly applied and enforcement only really occurs when there is a complaint or information breach. I believe that the best regulatory routes are perhaps industry benchmarks and standards such as the Visa AIS program. Enforcement is easier to control and standards can be set specific to the merchant or service provider risks.

Mahmood:

Well, we can try self-regulation for awhile and see whether that works. If that does not work, the government will have to be involved.

Arazi:

I believe that a balance between the two is ideal, such as a non-governmental committee in which business, finance, government, legal and technology professionals have rotating and time-limited seats. This committee should issue guidelines and have authorized inspectors conduct acceptance testing, similar to the way the ISO 9001 standard is implemented.

Q: What are the issues of secure payments that your organization has to deal with in its e-commerce environment?

Oliva:

The issues we have encountered are the same everyone has: fraud, mistakes in credit card account numbers, spoofing, etc. We receive credit card payments through a secured transfer with Verisign that filters most of these out. We make bank wire transfers through encrypted websites and transaction-specific Web cookies.

Naglost:

Maintaining privacy and security of credit card details has been Berkeley's primary concern. Where possible, Berkeley facilitates a direct transaction between the consumer and transaction gateway, ensuring that only the required parties have access to the credit card details. This solution, however, is not always possible—especially when working in our capacity as an intermediary e-commerce organization that manages order processing involving automated faxing of customer orders to a non-technology focused business. This process, although being as secure as any standard fax transmission, does raise additional privacy and security concerns that must be made clear to all parties.

Thompson:

Our dilemma is whether to use real-time transactions through a third-party vendor or provide the secure transactions through our own environment and process the credit cards manually. This is an issue when low transaction volumes are concerned, as most third-party vendors charge a monthly fee as well as a per-transaction fee.

Upadhyaya:

We have been quite lucky on this aspect. So far, we have not experienced any fraud or interception of payment details. So again, in terms of being proactive, we are reserving technical resources until the threat arises.

Arazi:

We have only dealt with EDI and credit-card based payments. With EDI, the financial framework is typically pre-arranged and thus there is less chance of fraud, and the "value" of any information obtained via unauthorized means is drastically diminished. Thus far, credit card payments have been accepted via secured HTTP forms and have been validated offline. Once validated, only the card type and last four digits are retained by the e-commerce systems, with the customer being prompted to either use the authorized card (without inputting its information again) or adding a new credit card, in which case the validation cycle repeats again. Only the actual payment server would store the full credit card information, and this information would always be encrypted on a highly secure server.

Thomson:

A primary task is the ongoing maintenance and evaluation or platforms and software to ensure that security patches are installed and known security holes closed. We constantly have to deal with nuisance attacks (port scans, small scan DOS, probing) and the subsequent blacklisting of IPs and/or ISP notification. We also have to devote R&D time to evaluate and adopt new industry standards and services such as VbV and SecureCode in order to meet our obligations to our financial partners and clients.

Mahmood:

Our biggest concern at UT-El Paso is client identity–making sure that the client cannot say that he did not order something when he indeed ordered that product or service...the security of the transactions while it is taking place.

Seen:

At Murchison, we operate an in-house credit card gateway and process the most online payments in the state, so it is vital for us to stay on top of security in respect to both sides of online payment processing. Foremost for us is ensuring that the gateway machine has a high level of availability and cannot be compromised in any way. The key is ensuring that we are vigilant in applying OS patches and provide proper firewall protection. From a front-end perspective, we ensure that transactions are only accepted from selected hosts, with proper authority, keys, certificates, etc., to negate any chance of spoofing or replay attacks.

Q: How does your organization deal with the issues of retaining e-commerce skilled personnel? What kind of attracting/retaining program(s) is in place for the proper staff?

Oliva:

Good question! We have had to train our staff in e-commerce security practices through a combination of online classes and on-the-job training. We recruited people eager to attend this training program and who had the proper system administration experience.

Intelligent Decisions LLC encourages staff members to continue their professional education through a combination of online and on-ground classes offered through commercial providers such as the IEEE, ISSA, The Chubb Institute, CISCO and the Northern Virginia Community College. Topics include data encryption, information assurance, firewall management and enhancement, network management, biometric authentication, "white hat" network security methods and GISRA (U.S. Government security standards). All staff members have most of these skills when hired, but they must constantly update them through seminars, classes, conferences and literature review.

Thompson:

Since being established in 1998, CECC has achieved the greatest success in attracting and retaining staff by drawing staff from the student population of the University of Ballarat and then supporting staff as they develop their e-commerce skills. The integration of and retention of staff engaged through more traditional recruitment methods has been less successful.

Thomson:

We are located in a market where there are a number of well-trained e-commerce employees (two universities and two colleges), predominately small technology companies (wage costs are therefore at the low end of the scale) and where quality of life is an important factor. As a result, there is always availability of staff at reasonable rates. Once we have employed and trained staff, we try and retain them by focusing on flexible work hours, a positive work environment, a team-based approached to work and ongoing training. Having said that though, finding good e-commerce security staff that don't have a criminal record is very difficult in our market!

Naglost:

Berkeley provides a number of incentives to retain its e-commerce skilled personnel. This involves flexible working schedules, payment mechanisms and company ownership. We utilize a number of key recruitment agencies, with whom we have formed strong partnerships, to attract quality staff. By providing flexibility regarding working location and remuneration, we have been able to achieve excellent overall retention.

Seen:

We have no formal programs in place. As a small organization, we have low staff turnover, so this is perhaps not as big a problem as in other enterprises. The downside is that there is a large chunk of institutional knowledge tied up in individual staff. Ongoing efforts to document and cross-train others helps to address this.

Arazi:

We do not have any special retention programs. We have not had to deal with high turnover of skilled IT personnel and as such did not develop any special retention programs. I have heard of problems in other organizations, but I feel those were endemic to those particular organizations and unrelated to the specific skills of the personnel in question. To retain good staff, I would recommend that the organization recognize their value, compensate them appropriately, work with them to solve problems (rather than lay blame for problems) and create an inquisitive work environment where each individual is able to pursue challenges and gain exposure to new ideas, technologies and processes, both within the organization and outside of it.

Upadhyaya:

It may sound tragic, but at the moment the UK e-commerce employment market is at a low. Staff say it is because they have jobs! We all have colleagues/friends that are unemployed at the moment. I'm sure this policy is short sighted and as soon as the market picks up, labor turnover will rise. I think when the time comes, a stake in the company is all that can really be offered.

Q: How does your organization deal with the issues of interoperability (maintaining reliable exchange of information with corporate information systems)?

Seen:

Generally, we do not deal with solutions that interface with corporate systems. Our target market is the SME sector, and as such, the solutions we offer are self-contained online storefronts. Theoretically, a customer needs no other software to operate their online business. In practice, however, our consumers tend to prefer exporting reports to third-party accounting software through an intermediate format such as CSV or XML.

Oliva:

As a small company, this is not a problem for us. However, for our large customers, this is a major issue. They use a combination of trusted system software and protocols, limited network access, and manually approved data transfers to move mission-critical data back and forth between field and corporate systems.

Thompson:

We do not have a great deal to do with this, as a vast majority of our business is online and we do not provide procurement services.

Upadhyaya:

When we embark on a data-sharing exercise, we normally send a consultant to meet with the IS department of the target company. And it is their task to come up with data transformation rules and compromises. We then get a sample data export or snapshot of the database to test our data mediation.

Thomson:

Part of our competitive advantage is the fact that our solutions bridge the gap between banking platforms (inflexible) and merchant financial systems. Our systems and solutions are all Web based and accessible using IS or Netscape, and we deploy a wide range of Java, C++ and CGI-based APIs to try and bridge as many platforms as possible. Our solutions are designed in such a way that they layer integration into the business system. All of our core products run on their own independent platforms and it is only the front and back end hooks that need to be customized to a specific system.

Arazi:

Interoperability is very important for A2i, yet we strive to maintain the mininum amount of copies of sensitive information. Pseudo-random IDs are used to transfer information between systems, with the recipient system then requesting the information from the secure server that holds the master copy of the information. If authorized, the information is sent to and used by the recipient server for the requisite task, after which the master server is updated, if applicable. In addition to ease of interoperability, this also provides high security, granularly regulated access and reduced likelihood of multiple copies of the same information being out of synchronization.

Mahmood:

My organization uses state-of-the-art technology in the area.

Naglost:

Berkeley deals with interoperability by utilizing industry standard information systems and architectures, and by creating systems and processes that facilitate simple integration between disparate systems and organizations. The benefits Berkeley IT realizes from these systems are wide and varied, and include the ability to obtain a single viewpoint of all company interactions with the customer/partner/competitor and to streamline the training and development overhead for new and existing employees and partners. Berkeley IT has also seen major improvements in the management of customer relations, as all disparate information is integrated and accessible to all employees.

Q: In your opinion, what are the technology limitations in assisting an organization in providing a total security environment in e-commerce systems?

Thompson:

These days there are not many technological limits in regard to e-commerce systems. Most problems arise from retaining skilled staff and other organizational issues. Systems need to be able to evolve and adapt to new security technologies as they become available.

Mahmood:

A lot of technologies including security technologies are still evolving.

Arazi:

And I would agree that there are no significant technological limitations. In addition to a large body of literature and reference works on the subject, there are sufficient numbers of personnel, software and hardware packages (including open-source ones) available. These resources should allow any organization that is capable of developing an e-commerce system to also integrate the requisite security programs, precautions and procedures. Thus, the question becomes one of motivation, since the ability is obviously there.

Oliva:

In terms of technology, data encryption is available and easy to use, and Internet transaction security has passed what happens in the non-Internet world (restaurants and retail stores, for example). The limitations are more investment cost and business process driven than anything else—if you have the money to spend, you can do a really good job at protecting yourself and your customers.

Thomson:

I find that as we develop mostly under Windows architectures and deal with clients with very high availability requirements, the single greatest limitation we see is the lack of "hot fixes" for security updates and patches. In most cases, systems must be taken offline in order to successfully apply a patch, and it can take 30 minutes of support time in order to properly reboot and re-establish a platform.

Seen:

Our biggest technical limitation is the reliance on vendors to provide timely patches to proprietary systems. It is often said that a chain is as strong as its weakest link. It makes it all the more difficult to reinforce the weak link when you need to rely on someone else to do this.

Naglost:

And wouldn't you agree that e-commerce security environments are usually limited by existing or inferior technology systems that are either not designed to allow additional security mechanisms to be implemented or have serious and inherent security flaws that cannot be rectified by external security mechanisms?

Upadhyaya:

Fundamentally, the problem is that the underlying technologies that we use can never be fully tested until they are unleashed on the real world. Who knows what security flaws will emerge in your operating system, Web server, database server, etc., until someone has tried to hack/exploit it? Because of this, we can never be totally safe, and if we wait for our software to become impenetrable, we will lose our competitive edge.

Q: In your opinion, what are the organizational and managerial limitations in providing a total security environment in an e-commerce system?

Naglost:

The major organizational and managerial limitation in providing a total security environment in an e-commerce system is the willingness of management to commit both funding and other resources to implement security systems—especially when this implementation results in major changes to both information system configuration and general business processes and practices. The organizational culture can also hamper this implementation if staff are unwilling to facilitate the process.

Thomson:

Lack of awareness of security issues is the single greatest problem that we see. In most cases business managers don't know what they don't know and therefore simply adopt solutions that they deem to be the most cost effective without understanding the full impact or risks of their decisions. At the moment there seem to be competitive certification processes for security specialists which further complicates the consulting and tendering process. We find that most companies are reactive rather than proactive with security measures. They undertake analysis and action after they have experienced a security breach and have done little in advance to prepare or plan for a security incident. Even after an assessment has been made, many firms fail to make the assessment process an ongoing one.

Upadhyaya:

I would add, fear and budget. On one end of the spectrum we fear the attack and what we don't understand. Then, ironically on the other, a budget forces us to buy non-industry standard protection devices such as poorly documented firewalls that lead to misunderstandings and even more security flaws.

Arazi:

Certainly, management is often interested in the "bottom line," i.e., the financial aspects of the business and the amount of profit generated by the e-commerce system. It is therefore best to build the case for a security policy and implementation as one that will have significant upside and a large downside if done wrong or not at all. Once the initial barriers have been cleared, it is best to explain why certain things happen, rather than just how, to the personnel that develop, implement, support, maintain and man-

age the e-commerce system. With this understanding, they will prove to be assets, often suggesting technical and managerial improvements to security policy.

Thompson:

But often the decision makers are not those with the operational knowledge of the options in relation to the security environment or in e-commerce systems.

By not having in-depth operational knowledge of e-commerce security, decision makers are handicapped when dealing with their own technical staff. Information technology by its nature changes so quickly that a person with management responsibilities would find it impossible to keep abreast of new developments and technologies. This creates an information gap between managers and technical staff, which in turn makes it hard for the decision makers to ensure that standards are being adhered to and that recommendations made by technical staff, are the correct ones. This is not a unique problem and is best dealt with by proper staff management and training.

Oliva:

My list of organizational limitations would include:

1. *Utilizing multiple systems requiring multiple passwords*
2. *Partial customer service capabilities due to segregated access to customer information*
3. *Changing passwords every seven to 10 days*
4. *Retraining on security procedures every 90 days*

Managerial limitations would include:

1. *Limited information access vs. trying to increase user productivity through open access*

2. *Need-to-know access management cost vs. trusting all users*
3. *Funding user education and awareness classes on a very frequent basis*
4. *Funding investments in hardware and software without revenue return*
5. *Funding real-time response teams for hacker and virus attacks*

Mahmood:

I think we all need to admit, though, that there is no such thing as total security environment. If hackers want to break into systems, they can find a way. We must try our best to stay ahead of the hackers.

Q: During the past several years, many organizations have been questioning the actual dollar returns on their investments on information technology-related security programs. In your opinion, how can organizations measure returns in their investment in e-commerce security-related programs? How has your organization accomplished this goal?

Naglost:

Information technology-related security programs should never be measured in terms of actual dollar returns. Just like physical office security or personal security, information technology-related security programs are a necessary business operating expense that limits the possibility of information technology compromise or destruction. It is this information technology compromise or destruction that may have catastrophic "actual dollar" consequences to the organization if its information technology is left unprotected. Organizations can measure the returns in their investment in e-commerce security-related programs by gathering feedback from their existing customer base and internal employee base on their perception and real experience of the organization's e-commerce security. Berkeley IT has accomplished this goal by directly and regularly surveying both our employees and customers on our existing e-commerce security-related programs. We also involve industry security experts to formally assess our security programs. From this feedback we can gauge our successes and failures, and can work toward improving the overall level of security across the entire organization. If your customers, employees and industry experts are satisfied with your e-commerce security and you are not experiencing security compromises, then chances are you are doing it the right way!

Oliva:

At Intelligent Decisions LLC, we calculate the return on investment two ways:

(a) The intangibles are: direct operating costs not incurred due to having IT security processes and technology in place that stop DoS (denial of service) attacks, reduce the impact of worms and viruses, and unauthorized access or theft of data through the network. We compute our "savings" based on work activities we do not perform.

(b) The direct returns comprise new business, return business and referrals due to our success in keeping our information safe and secure, and satisfying contractual and ethical obligations involving our customers' data. We accomplish e-commerce security through a variety of approaches including digital certificates, trusted access authentications and firewalls. These capa-

bilities, in turn, allow us to build a business based on successful performance, trust and mature, software tools.

Thompson:

CECC has minimized the dollars spent on e-commerce by implementing freely available OpenSource solutions wherever possible. OpenSource software provides small organizations with cost-effective e-commerce security without compromising the robustness of the solution achieved. Measuring the returns on investment is vastly easier when the initial investment is only in terms of staff time. One breach of security could cost an organization its reputation and have a massive impact on revenues. CECC therefore believes this staff time investment is critical.

Arazi:

Rather than looking at positive numbers, an organization would have to evaluate how much it stands to lose if no investments are made in IT security. When considering the purchase of an alarm system for a vehicle or a home, one does not calculate the profit but rather the potential losses without the alarm in place. Similarly, it is difficult, if not impossible, to quantify an organization's net gains or losses, but a consensus exists that news stories about an organization being "hacked into" and customer data or credit cards stolen are certainly undesired and often very damaging. We have accomplished this goal by attempting to measure the negative effects of not implementing proper security policies, measures and procedures. The results spoke for themselves.

Thomson:

We view risk management and security as an "insurance policy." It is something that cannot easily be quantified in terms of a "return on investment" but can be justified if you consider the cost vs. analysis. The negative publicity and loss of confidence associated with a security incident are more than offset by the cost vs. benefit analysis. In my city, a local grocery store recently experienced their own version of a comparable incident. The store discovered that they had an employee that tested positive for Hepatitis A. The grocery store had a contingency plan in place, and they quickly issued a press release outlining the concern and announcing that they had destroyed the produce that the staff member was responsible for handling. A plan for testing all staff was also announced. The net result was a very positive consumer response and media exposure for a relatively small investment. Had they tried to conceal the incident or not been prepared, the results would have been significantly different!

Upadhyaya:

That's right, because in financial terms, you need to look at your security as opportunity cost. Firstly, without any strategic security planning/implementation, your organization will waste time, energy and resources surviving in a reactive state, that is draining and destroying the inventiveness of your IT department. This will have an effect on your competitive advantage, as your department spends more time dealing with issues as they arise, rather than having time to consider R&D or improving existing systems. Thus, your organization will continue to hobble along with costly security inefficiencies that could have been eliminated with investment in e-commerce-related problems. We were like this two to three years ago—we looked at all the man hours spent being reactive to security issues. We then multiplied that time by the chargeable hourly rate for each member involved in resolving the problems, and then time wasted by other staff in the organization due to outages. We then compared that with the cost of e-commerce security programs! All this was without looking at how the outages were affecting our image with clients and potential clients. The bottom line is that the cost of not having a secure e-commerce-related program is much higher than having one.

Mahmood:

Thus, organizations should focus on doing a good job in measuring qualitative benefits?

Seen:

Absolutely. Simply put, measuring the return is difficult, as it usually involves measuring losses which have not occurred. What we have done is look at past issues which have occurred and assessed what our exposure would have been if security measures had not been in place. For example, with the Slammer worm, we determined that our security policy saved us from a potential loss of bandwidth and excess traffic charges, and allowed us to maintain service. In this case, there were two main parts of our policy which came into play: proactive security, in the form of filtering network traffic to the SQL server, and reactive security, in the timely application of patches for SQL Server when they were first released. Certainly, the time and effort invested in applying patches and configuring our firewall was, at a conservative estimate, one fifth of the potential loss if our SQL server had been infected by the worm.

Q: In your opinion, what role should government play in protecting consumer/client rights and privacy in e-commerce transactions?

Naglost:

Government should legislate and enforce regulations that unequivocally protect consumer rights and privacy in e-commerce transactions. It is the responsibility of all government organizations at both the federal and international level to work together to ensure regulations are implemented and enforced, and are not affected by national boundaries.

Mahmood:

That's right. You have to admit, the laws are a little behind in the e-commerce area.

Thomson:

I disagree about government. I think that the government should not be playing a major role in regulating e-commerce transactions. The legislation in Canada and the UK that protects consumer privacy and the confidentiality of information on a broad scale (not just e-commerce) is a good idea, but attempting to regulate individual businesses will become manpower intensive and not yield the results they want. At this point I believe that industry regulation is the best measure.

Thompson:

But governments play an important role as an information provider and may be able to effectively disseminate information that will protect consumer/client rights in relation to privacy in e-commerce transactions.

Upadhyaya:

Similar to an earlier question, the government should be involved in a way that doesn't infringe on the information freedoms of the Internet, but not be powerless to help the consumers. There already are models forming around the globe at the moment–the information regime of China, America's recent laws regarding national security and digital information, and the UK's attempt to please homegrown surfers. Each a different angle, but none to an overwhelming success. Fifty percent of consumers still feel uneasy about e-commerce transactions.

Seen:

I believe that the role government should play should be equivalent to the role governments play in traditional commerce transactions. There are many rules and regulations that can equally be applied to e-commerce transactions as there are to those conducted by phone, fax or e-mail. Increasingly, this situation is becoming more complex because of the ability for the average user to import goods simply by making a purchase online. Government cannot be expected to mandate particular security measures, such as 128-bit SSL sessions, etc. It is up to the marketplace to regulate this itself and refuse to deal with those businesses which do not comply.

Arazi:

I believe that capitalist forces will eventually dictate this rather than any attempt of government intervention. If a certain business can provide a competitive (presumed or real) advantage over another as a result of better protections, that business shall prevail. The customers will decide.

Oliva:

This will continue to be a battleground for years to come, due to the difficult balancing of business access and financial commerce against customer privacy rights. The government should propose standards that form a "trust agreement" between businesses and customers that provide these minimal levels of bilateral security:

1. Identity authentication of both the seller (company) and customer. Both parties must have confidence they know who the other is.

2. Information that is exchanged must be safe from theft or corruption by unauthorized parties while in transit (probably through the Internet) between them.

3. After receipt of data between parties, business process safeguards must protect customer information from unintended disclosure to third parties or unintended use.

4. The last step, but hard to accept, is for businesses to pay customers for the use of their information through "opt-in" choices. However, this creates yet another set of security questions about keeping those identities safe from unauthorized use.

Section II: Challenges, Solutions and Future Issues Facing E-Commerce Security

Q: In your opinion, what are 3-5 current challenges facing practicing e-commerce professionals in modern organizations in dealing with e-commerce security?

Naglost:

1. *The ever-increasing number of hackers who have the skill to penetrate e-commerce security systems.*
2. *The growing number of new applications that utilize unique security methods, each with its own inherent security flaws.*
3. *The requirement to integrate with external systems that often use disparate architectures and information systems.*

Oliva

1. *Obtaining management support and respect of their skills and importance to the business.*
2. *Continuously investing in professional education required to stay ahead of the criminal community.*
3. *Educating users and obtaining their cooperation in basic security practices such as not sharing passwords, changing passwords and blabbing about private information in public.*
4. *Walking a fine line between "locking out" all users of company data vs. keeping systems open to customers, suppliers, and employees by utilizing the inherent security abilities of automated systems.*
5. *Complying with ever-changing and always conflicting governmental laws and regulations.*

Thompson:

1. *Lack of client expertise — clients are often unable to adequately assess alternatives based on e-commerce security, so decisions are made based on factors such as price, company location, etc.*
2. *Cost*
3. *Pace of change and degree of redundancy*

Arazi:

1. *Security is often regarded as a non-profit-generating expense and is therefore not given sufficient funds.*
2. *Security is often taken for granted. It is almost totally ignored until the very end of the specification and implementation phases, after which it is added as an afterthought, if at all.*
3. *Lack of understanding that securing an e-commerce system is an ongoing process that must start with the initial specification and continue through every day the production e-commerce system is functional.*

Upadhyaya:

1. *In recent years, e-commerce has attracted interest from businesses and consumers alike which has caused growth in the number of B2C transactions. Even so, I doubt that e-commerce will reach its full potential until customers perceive that the risks of doing business electronically have been reduced to a tolerable balance.*
2. *Consumers may have justifiable concerns about control, authorization, confidentiality, transaction integrity and anonymity.*

Thomson:

1. *The nature and type of threat is constantly changing, and it takes considerable time and effort to identify and address the risks.*

2. *Businesses are becoming more and more dependant on their information systems, and the impact of any potential attack or outage continues to increase.*
3. *Technology solutions are becoming increasingly complex, and identifying and resolving security risks at the application level is becoming more difficult.*

Mahmood:

1. *The challenge to stay ahead of the hackers*
2. *The need to deal with integrity threats and necessity threats (DOS)*
3. *Dealing with Web server threats and database threats*

Seen:

The biggest challenge is still education. Educating non-technical clients about what measures must be taken to make a good-faith effort to protect the consumer. Also, consumers need to be educated about where to put their credit card number and where not to put it.

Other challenges include the frightening prospects of systematic identity theft. Given that e-commerce sites collect private and often confidential client data, they can present a prime target for skilled groups who aim to perpetrate these crimes. In the case that an attack is motivated by the opportunity to steal data rather than gain a financial advantage, it is difficult to detect when such an attack has occurred; there are no balance sheets recording who has accessed what data.

Q: What are 3-5 solutions that you can recommend to practicing e-commerce professionals in managing e-commerce security effectively?

Naglost:

1. *Adopt a continuous learning approach to e-commerce security including consistent, periodic investigation and research of online, lecture and periodical-based material to ensure maximum exposure to all current and emerging issues.*
2. *Plan and document all systems—especially the overall security architecture. This will ensure that all areas of security are covered. Discuss the architecture in simple terms with management and gain top-level support for the overall plan.*
3. *Implement reliable, tested and industry standard solutions where possible and ensure sufficient business processes are developed to facilitate the overall security objectives.*
4. *Use e-commerce security specialists to assist with the development, implementation and testing of your security systems to ensure complete and comprehensive solutions.*

Oliva:

1. *Know your users, customers and suppliers in order to be able to isolate strange activities without delay.*
2. *Plan for the worst-case security breech scenario and implement proactive processes and tools to stop it from happening, if possible.*
3. *Train management about what is possible and impossible for you to do. They need to know how they can help you succeed.*
4. *Be realistic when developing budgets—don't ask for the sky but don't ask for tools that can't protect the company's data assets.*
5. *Use common sense in enforcing security regulations and restrictions.*

Thompson:

1. *Ensure that technical issues are explained clearly*
2. *Provide links to external information on security issues wherever appropriate*
3. *Partner with like organizations to reduce the costs, share knowledge and aggregate to secure more sustainable outcomes*

Arazi:

1. *Increased awareness in all levels of the organization will greatly boost security. Overall security will increase as different departments see that IT, business and management have "bought in" to the need for security and are diligently working on solutions.*
2. *Decision makers, designers and implementers within an organization must be educated to think in a security-conscious manner when performing their duties. Once this happens, business decisions will drive technological specifications that will evolve into secure implementations of e-commerce systems, to the benefit of all.*
3. *"Hands-on" management, that is, security professionals must rely on themselves as well as on others while recommending policy and implementing it. Much like doctors or mechanics, one must practice and personally experience the subject rather than just read theoretical material about it.*

Upadhyaya:

1. *Thin-clients offer greater reliability and increased security. All data and applications held centrally and not on desktop...greater security and control of user access to network, applications and data...no risk to infect desktop with viruses or "snooping" programs, and high server protection against hacking and intrusion.*

2. *Self-hosting. What I mean is, do not share your server with other companies. It is common sense: if you have the infrastructure to be able to provide e-commerce solutions, then spend your budget on your own box.*
3. *Encrypt your databases, be more proactive and protect your client's data. Implement detailed security policies that only grant access to certain applications and IP addresses.*

Thomson:

1. *Adopt industry best practices on the protection of confidential information.*
2. *Outsource security monitoring and risk management if it is not a core business competency and internal resources are lacking.*
3. *Treat security and risk management as an ongoing task rather than a one-time evaluation or certification.*
4. *Don't overlook the weakest link in the organization (typically staff) when developing a security policy. There is no point building a steel door for a tent!*
4. *PGP signing e-mail...slow uptake, very small early adopter market share. The biggest problem in our industry with this is that we all use secure thin clients to gain access/register with our online banks/e-commerce retailer. Then our passwords are sent back to us in a plain-text e-mail. Do not do that—it is very silly and pointless. Never send any sensitive information in plain text.*

Mahmood:

1. *Digital certificates*
2. *Using SSL*
3. *Using firewalls*

Seen:

1. *Be aware of developments in security, not just in e-commerce, but in computing generally. Keep abreast of the security flaws that are being exploited, evaluate vulnerability whenever necessary and act immediately. There are many options for security training with a range of organizations that can offer formal training in security and loss-prevention techniques. Alongside these courses, there is plenty of scope for exploring usenet groups and other forums in which security-oriented discourse takes place.*

   Vendor-independent mailing lists like CERT Advisories or Bugtraq provide an objective method of tracking developments in security issues and can be a useful addition to mailing lists and announcements published by software and OS vendors/developers. This is an industry-wide issue, it is important to track current security issues in many fields of IT, but the high exposure of e-commerce sites makes this even more crucial.
2. *Think critically. Actively try to pick holes in the systems under development. Take the "black hat" approach and try and spot where and how you would do the most damage to a system.*
3. *Know how to respond if the worst happens. Know who must be contacted in the event that the worst occurs. Depending on the compromise, this may include banks, financial institutions, merchant service providers, and of course, clients and consumers directly affected. Develop an internal "damage control" plan and review this regularly in light of new security threats. Just going through this process will help to keep security foremost in everyone's minds.*

Q: What are 3-5 future challenges that will be facing practicing e-commerce professionals in modern organizations in dealing with e-commerce security?

Naglost:

1. *The security threats will not only come from external sources but as local staff become more technologically advanced, internal threats will become a major factor.*
2. *The integration of multiple organizations' e-commerce systems will become commonplace, and the ability to ensure a high level of security within this multi-organizational system will be paramount.*
3. *E-commerce security in the mobile arena will increase as m-commerce increases rapidly over the next decade. E-commerce professionals will have to be able to understand m-commerce security requirements and integrate these into their existing e-commerce security environments.*

Oliva:

1. *Blocking increasingly sophisticated attacks from organized international criminal elements (cyber criminals and terrorists).*
2. *Overlapping and conflicting laws and regulations from international, federal and state governments making it difficult to not be in violation of someone's laws all of the time.*
3. *Minimizing liability for financial damages when an accidental breech of information or trust occurs—lawyers will want to extract maximum penalties from all parties, even if an unavoidable situation happened.*
4. *Meeting the always-increasing expectations of customers, staff, management and suppliers for "airtight" security, all the time, for free.*

Thompson:

1. *E-commerce professionals acting in isolation will find it increasingly difficult to keep up to date and to ensure that their organization is adequately prepared to deal with all issues associated with e-commerce security.*
2. *While e-commerce security will be a continuing issue, it may become more difficult for e-commerce professionals to convince their organizations to invest, to address e-commerce security issues.*
3. *The issue of public perception will continue to impact on perceptions of e-commerce security. In the absence of some coordinated effort by e-commerce professionals, media organizations are likely to continue to focus on "horror stories" rather than promoting stories of organizations successfully dealing with e-commerce in terms of security or other areas.*

Arazi:

1. *Improved hacking techniques – Cyber-crime is on the rise, and thus cyber-criminals are constantly evolving to employ better penetration and attack methods. This is especially true as e-commerce systems proliferate, thus increasing the potential "reward" for a successful breach.*
2. *Increased financial pressure – As businesses, especially technology-related ones, become more concerned with the financial "bottom line," they will try to cut costs wherever possible. This may lead to a situation where e-commerce security will not be allocated the resources required for a proper implementation and ongoing maintenance and upgrades, or to a scenario where security will be integrated into an organizational division that may not know enough about the threats.*

3. *Increased bureaucracy – As legislation and regulation of e-commerce environments increase, situations may arise where implementations of security systems may, temporarily or permanently, be affected by laws and other rules and regulations. This may cause certain implementations to become very cumbersome.*

Upadhyaya:

1. *Emerging mobile technologies and the security implications involved such as reduced bandwidth and processing overhead.*
2. *Dissemination of technologies within the information militia (the cracker/hacker communities), for example distributed attacks like DoS and brute force key cracks.*
3. *Complying with future regulations, bringing your current information/e-commerce solution in line with legislation.*
4. *The need to encapsulate future communication standards such as building "future-proof" security measures.*
5. *Embracing new and old forms of private data encryption, e.g., PGP, etc.*

Thomson:

My future challenges are largely the same as the current challenges with the addition of: Government regulation creating further paperwork and workload for businesses who already adopt best practices.

Mahmood:

1. *To help customers understand the security protocol*
2. *To make sure that customers feel safe doing business online*
3. *To make it as easy as possible to shop online*

Seen:

1. *In my company, we are working toward developing decentralized, peer-to-peer e-commerce systems. In the P2P realm, most of the conventional thinking regarding e-commerce security gets thrown out the window. Issues of mutual trustworthiness, non-repudiation, etc., are challenges that professionals in the sector must face. Given the tremendous promise for P2P in e-commerce, I have no doubts that we will see increasing coverage of these issues in the future.*
2. *Another challenge we face is the "commoditization" of e-commerce solutions. When a consumer can walk into a department store and buy an "e-commerce in a box" solution, it devalues the hard work and effort that goes into developing a tailored, secure product. E-commerce professionals need to value-add in every way possible if they are to continue to justify their existence for all but the largest of projects.*
3. *Finally, I can see connected mobile devices experiencing a resurgence along with the spread of true 3G capability in the cellular network. Mobile and ubiquitous computing devices will offer the potential for anywhere, anytime e-commerce transactions. I can see many challenges and opportunities in tailoring high-availability e-commerce systems to this marketplace.*

Appendix A

Panel Member Profiles

Mark Naglost
Berkeley Information Technology
Managing Director

Mark Naglost is the Managing Director of Berkeley Information Technology Pty. Ltd. (www.berkeleyit.com), a leading Australian company specialising in Customer Relationship Management and Internet services. He has filled this role for four years and has also filled other senior management and board positions including the CTO of OzStudios Pty. Ltd.

He has a Masters of Science in Information Technology from the University of New South Wales and is a full member of the Australian Computer Society. He also holds a number of industry certifications, including Microsoft Certified Systems Engineer. Mr. Naglost has almost 10 years of experience within the IT industry and specialises in Internet and intranet systems with a focus on system integrity and security. He can be contacted at mark@berkeleyit.com.

- *Company's main products or services*

  Customer Relationship Management and Internet Services.

- *Number of employees*

  Ten to 15 staff including a mixture of full-time, part-time and field-expert consultants. Berkeley IT is a reactive organisation that can grow to accommodate almost any project or requirement via its comprehensive and cohesive partner network.

Matan Arazi
A2i, Inc.
Director of IT

An Israeli native that grew up in the U.S. and Japan, Matan Arazi spent three years in the Israeli army's elite intelligence unit, where he analyzed and accumulated diverse knowledge of current computer, communication and IS standards, practices and technologies. He then became CTO of KenyonNet, Israeli's first full-scale online shopping mall. Following this, Arazi joined Conduct Software Technologies. Next, he spearheaded the development of Israel's online National Photography Archive. He then accepted a position at A2i in Los Angeles, and is currently in charge of designing and overseeing the deployment of A2i's leading-edge accelerated cross-media catalog publishing and e-commerce solution, where he pays particular attention to deployment and security topics.

- *Company's main products or services:*

  A2i has developed xCat, a scalable system for enterprise-wide content management and catalog publishing. A2i's cross-media publishing system is database driven, blazingly fast and supports up to millions of products. The A2i xCat system is a new class of application software for enterprise-wide content management, catalog publishing and system integration.

- *Number of employees:*

  Approximately 100.

Lawrence Oliva
Intelligent Decisions LLC
President

Lawrence Oliva is the President of Intelligent Decisions LLC, an IT services company providing secured e-commerce, database and knowledge management services. Mr. Oliva has been an executive in the technology industry for more than 20 years, including senior management roles at Sun Microsystems and Dell Computer. As a certified program manager, he has led global project teams, building and implementing e-commerce and biometric security systems for large and small companies. As a senior consultant, he has selected and justified investments in e-commerce software and systems, and developed business and pricing models obtaining user-paid revenues from online transactions. He holds an MBA and a bachelor's degree in Organizational Behavior. Mr. Oliva is a noted author of technology subjects and is a senior editor for the Information Resources Management Journal *and other Idea Group Inc. publications.*

- *Company's main products or services:*
  E-Commerce, Knowledge Management and Information Security Consulting Services.

- *Number of employees:*
  Intelligent Decisions LLC has a staff of eight people, all with expertise in IS, network management and government security standards.

Helen Thompson
Centre for Electronic Commerce & Communications (CECC) Manager

---

Helen Thompson manages the Centre for Electronic Commerce & Communications at the University of Ballarat, Victoria, Australia. The organisation makes scaleable and customizable applications to meet the online service needs of its diverse client base. These applications include database development, registration systems, searchable directories, online project systems, event calendar tools, news and content tools, web linkage libraries, and help desk facilities. Her area of research interest is success and decline in regional and rural communities. More specifically, Helen is currently exploring the benefits that communities–as represented by local government, businesses and community groups– secure through e-commerce and community informatics initiatives.

- *Company's main products or services:*
  Customer Relationship Management and Internet Services.

- *Number of employees:*
  230 academic staff, 200 tertiary and further education teaching staff, 380 general staff.

Acknowledgment

The contribution of Andrew McLeod, the technical manager for the Centre for Electronic Commerce and Communications, should be acknowledged for his assistance in preparing for the panel.

Mayur Upadhyaya
Magic Lantern Productions
Information Systems Developer

Mayur Upadhyaya is an Information Systems Developer for Magic Lantern Productions, where he is responsible for working with designing back ends of Web applications, DBA, e-commerce security, system specification, XML/XSLT, Oracle and Java, infrastructure, and the development of several content management systems for both browser and desktop. His past professional experience includes working as an IT Communications Assistant and Computer Coordinator, where he installed and set-up remote access servers, developed VBA solutions for accounts departments and evaluated redundant hardware for maximum efficiency. He received a lower second class honors BSc in Business Information Technology.

- *Company's main products or services:*
  Content Management Systems for the Web, DVC, broadband, iTV, games and digital video. Provide consultancy for information technology, project management and strategic planning.

- *Number of employees:*
  Twenty-five.

Craig Thomson
Beanstream Internet Commerce Inc.
President

Technological competence is central to the health and future success of a modern enterprise. With a proven record of leadership combined with 12 years of success as a high-tech CEO, Craig Thomson has the ideal background to lead Beanstream. As Founder and President of Paradon, Mr. Thomson has been awarded the prestigious "Entrepreneur of the Year Award" (1994) in manufacturing, wholesale and distribution for Pacific Canada by Ernst and Young. His company was listed on Profit Magazine's *Profit 100 list as one of the fastest growing Canadian companies for 1996. He is a member of the University of Victoria Board of Advisors to the Faculty of Business and has been involved with YEO since its inception. Mr. Thomson received his degree in Computer Engineering at the Royal Military College of Canada. Prior to Paradon he spent eight years as a Naval officer working in combat systems and specializing in the areas of data security, encryption and digital communications.*

- *Company's main products or services:*

  Beanstream is a financial and authentication services company. We provide a wide range of cost effective and universally available Internet accessibility tools and applications to enable payment processing and risk management for businesses and their consumers.

- *Number of employees:*

  Fewer than 200.

M. Adam Mahmood
University of Texas at El Paso
Mayfield Chaired Professor

Mo Adam Mahmood is a Professor of Information Systems and the Ellis and Susan Mayfield Professor in Business Administration at the University of Texas at El Paso. His research interests include information technology in support of superior organizational strategic and economic performance, group decision support systems, software engineering, and the utilization of information technology for national and international competitiveness. Dr. Mahmood has published in many scholarly publications, including MIS Quarterly, Journal of Management Information Systems, Decision Sciences, European Journal of Information Systems, Expert Systems with Applications, Information and Management, Journal of Computer-Based Instruction, Information Resources Management Journal, Journal of Systems Management, Data Base, International Journal of Policy and Management, *and others. He has also edited and published a book in the information technology investment and performance area. He is currently serving as the Editor-in-Charge of the* Journal of End User Computing and Security.

- *Company's main products or services:*
  Higher Education.

- *Number of employees:*
  Between 2,000 and 4,999.

Warren Seen
Murchison E-Commerce Programmer

Since graduating from the University of Tasmania, Australia, with a First-Class Honours degree in Computing, Warren has worked for Murchison E-Commerce as a Software Engineer. Trained primarily in C++ and Java programming, his current research and development work focuses on exploring the commercial possibilities and security implications of peer-to-peer networks for e-commerce in vertical marketplaces.

Warren's other computing interests include networking and security systems, Java development, open source business models and mobile computing devices. In his spare time, Warren manages to spend too much time with computers and not enough time outside enjoying the wilderness of Tasmania, although he promises to change that soon.

- *Company's main products or services:*

  E-Commerce Strategy, Design, Implemenation, Web Services Development, Content Management Engines.

- *Number of employees:*

  Approximately 10 permanent staff.

Section III

Expert Writings

Chapter III

How One Niche Player in the Internet Security Field Fulfills an Important Role

Troy J. Strader, PhD
IS Department
Drake University, USA

* *co-authors: Daniel M. Norris, Philip A. Houle, Charles B. Shrader*
(please see pg. 87 for biographies)

Biography

Troy J. Strader is Associate Professor of Information Systems at Drake University. He received his Ph.D. in information systems from the University of Illinois at Urbana-Champaign in 1997. Dr. Strader has co-edited two books, the Handbook on Electronic Commerce *and* Mobile Commerce: Technology, Theory and Applications *and his research appears in* Communications of the ACM, International Journal of Electronic Commerce, Decision Support Systems, Electronic Commerce Research, Electronic Markets, Internet Research, Quarterly Journal of Electronic Commerce, Annals of Case on Information Technology *and other academic and practitioner journals. He has work experience as a computer programmer and systems analyst.*

Executive Summary

This chapter examines an entrepreneurial effort to provide products in the Internet security marketplace. The specific focus is on a company named Palisade Systems, which is now faced with questions regarding their future business direction in this field (Mahanti et al., 2004). Current questions include how to take advantage of recent legislation regarding privacy and computer security, and the general increase in awareness of the need for security in the Internet and in related networks. In this chapter we discuss the Internet security marketplace, recent legislation and the creation of new opportunities for marketing Internet security products, and how Palisade's products may match these opportunities.

The Internet Security Marketplace

With most technology companies caught in the doldrums due to a flagging economy and waning consumer confidence, an intense new emphasis on security issues has been an opportunity for others. Analysts agree that of all the technology sectors, security companies are faring best amid the general high-tech slowdown as companies take measures to secure the services and information they provide. The U.S. market for managed information security services is expected to reach $2.2 billion by 2005, up from $720 million in 2001. This represents a compound annual growth rate of 25.4 percent (McGee, 2002).

The government has long been tracking intrusions into the computer networks of the Pentagon and other government agencies, as well as attacks on private universities and research labs. The tragic events of September 11, 2001, played a pivotal role in bringing the issue of computer security to the forefront. In anticipation of future cyber attacks and other forms of terrorist attacks, the Department of Homeland Security was created which has led to spending towards implementation of more secure information systems in both private and public sectors. Legislation such as the Digital Millennium Copyright Act (DMCA), Graham Leach Bliley Act (GLB), and Health Insurance Portability and Accountability Act (HIPAA) all play an important role in the push for implementation of security measures in industry.

Additionally, on December 15, 2000, Congress passed HR 4577, the Children's Internet Protection Act (CIPA), which requires schools and libraries that receive certain federal funds to filter Internet content. The act requires that schools and libraries implement technology protection measures to prevent young students from viewing websites that are obscene, harmful or that contain pornography. The Act also stipulates that schools and libraries create an Acceptable Use Policy (AUP) that outlines what materials are not appropriate for minors to view. The Internet safety policy must address the safety and security of minors when using electronic means of communication, and must protect minors in terms of the use and dissemination of personal identification information.

The Internet security market is highly fragmented with frequent acquisitions and occasional company failures. Security problems are constantly changing and the technology has to constantly evolve to match it. Therefore, there has been a trend towards increasing the speed of introduction of new security products. There has also been a gradual trend towards bundling of security products or other software packages affording some consolidation in the appliance solutions that offer simpler, more efficient and higher ROI solutions. This has also led to the reduction of the number of vendors.

The network security market can be broadly divided into the following products: firewalls, intrusion detection tools, content filtering of Internet packets, and security scanning and reporting of network traffic. There are many companies that are producing Internet security products. Some are small scale like Palisade Systems, and some are large companies like Symantec and Computer Associates which have commanded the major market share. As in many of the computer technology areas, it is the smaller players who lead their fields in innovation and product development.

Internet Security Market Opportunities and Palisade Systems' Product Line

Established in 1996, Palisade Systems is a developer of Internet security products. The company's mission is to help client organizations manage and defend both internal and Internet-based networks from benign network utilization issues and malicious electronic attack or sabotage. It is the goal of their solutions to ensure user productivity, preserve bandwidth, reduce liability issues and increase overall network security.

ScreenDoor

As previously mentioned, CIPA requires schools and libraries to certify that they are either in compliance with filtering requirements, or that they are in the process of becoming compliant by evaluating blocking software. Palisade has addressed this opportunity with ScreenDoor, their first product introduced in 1997, which is a Web filter and has the flexibility that allows users to choose sites that are blocked by categories, such as shopping, travel, hate speech, sports and stock trading. ScreenDoor monitors and blocks Internet access by protocol, port and server address using Palisade's patented core technology. ScreenDoor's Screened Out list is easily editable. One can easily add sites to, monitor or block categories or override sites currently blocked by the list. ScreenDoor helps manage productivity, bandwidth and liability issues associated with Web surfing by filtering inappropriate Web sites. Palisade's major competitor in this market, Net Nanny, can only be used to block sites on individual computers (Symantech, 2002). Palisade products, however, are able to manage thousands of computers from one location. Therefore, Palisade can cater to schools and libraries that need to block sites on all computers accessed by students.

The explosive growth of Internet file sharing programs also poses a unique problem for both parents and their children. Teenagers who use programs like Napster, Gnutella and Morpheus to search for music could be exposed to objectionable material. The most popular parental filter programs like Net Nanny and SurfControl's CyberPatrol cannot block access to pornographic material obtained through file sharing programs. There is an urgent need to address this issue in a different way.

PacketHound

The Cyberterrorism Preparedness Act of 2002 aims to create a set of best computer security practices for the government. The bill has not yet been passed, but there is a good chance that it will pass. If this law is approved, companies will be required to follow the prescribed best practices. Even when the company does not do business directly with government, they still may be working for one that does. As a result, the company might find itself being required to use a government-approved firewall or intrusion detection system, or an improved network management system. In addition, there may be a need to enhance authentication. Companies will also find that there will be auditing and reporting requirements to demonstrate compliance. From Palisade's perspective this act is a potential boon in lean economic times. The bill will make available grant money for research into various aspects of security to universities, who in turn can then work with private companies like Palisade to come up with new approaches to address these issues (Webdesk, 2002). Secondly, the bill also includes tax incentives, such as an accelerated deprecation for the purchase of computer software. This additional tax benefit for the purchase of IT and software products will provide incentives for enterprises to purchase such products and could positively impact Palisade's sales. Finally, it will require more companies to comply with security measures which open up a larger customer base for Palisade. If enacted, this law would open up a bigger market for Palisade's PacketHound that can block file-sharing applications and therefore address problems dealing with security, legal liability or bandwidth hogging. PacketHound is a network appliance that allows system administrators to block, monitor, log, or throttle LAN access to an expandable list of unproductive or potentially dangerous protocols and applications. PacketHound protects the organization by managing network traffic using a flexible rule set. Companies can block an entire network from accessing certain protocols or applications, block all access except in one computer lab, or block on a machine-by-machine basis. They can also use time-based rules to block access during critical hours but allow its use at other times, and block based on network load or the load of specific applications.

In addition to ScreenDoor and PacketHound, Palisade offers other technology components/products addressing various aspects of the security market ranging from intrusion detectors to network monitors. Most companies try to stop Web surfing with either an Acceptable Use Policy (AUP) or by installing firewalls or proxy servers. Although AUP can be set, it is extremely difficult to monitor and enforce these policies. When proxy servers and firewalls are employed, most of the network traffic has to pass through these devices which can lead to performance degradation. The danger is not just limited to performance degradation, but to the overall security of the enterprise. For example, most personal firewalls work by having a preset rules database that has a listing of trusted applications that it will allow access to and from the computer. This approach relies on someone knowledgeable about the particular product setting the appropriate rules which might not be the case for every instance of the product.

FireBlock

FireBlock addresses the network level access to services and systems on internal networks. It leverages Palisade's patented passive network management technology to not only track activity but also to enforce policy on the internal network. It operates at the network transport layer to track and control network activity based on source, destination and protocol employed. By limiting the internal network level access to only those services and machines actually required, FireBlock proactively compartmentalizes the network limiting unnecessary capabilities that enable illegitimate network activity. It also provides excellent information regarding network-level activity. It provides administration with information about incoming WAN (wide area network), VPN (virtual private network) or open firewall port traffic. FireBlock requires minimal training for use and little on-going maintenance.

SmokeDetector

No matter what kind of firewall, intrusion detection system or authentication security tools is in place, SmokeDetector can add another valuable layer of protection. SmokeDetector is a network appliance that detects intrusion activity in a network. It does not rely only on attack signatures, but acts proactively to protect the network without giving false positives or needing to continually update libraries of signatures. SmokeDetector is used to disguise critical servers and detect those trying to access them inappropriately. It is placed on a network as a "decoy" to mimic the organization's important servers in order to confuse and delay hackers. While delaying a hacker from accessing assets, SmokeDetector also captures and logs all information communicated during the session and sends an immediate e-mail warning to the administrator. By the time the hacker determines he/she is accessing a fake server, the administrator already has all the needed information and has locked down the real assets. SmokeDetector also captures important information from the hacker and sends it to the administrator. The information contains the date and time of the attempt, the IP address of the emulation (or "fake server") being accessed, the IP address of the person communicating with the SmokeDetector and a number indicating the "alert level," which is used to help the administrator gauge the severity of the attack.

FireMarshal

Internal security breaches account for nearly 80% of all violations reported by businesses and government organizations (Palisade Systems, 2003). For years, companies have been focusing on how to protect from the outside threat, while ignoring what their own employees and trusted partners are doing on their networks. FireMarshal utilizes a role-based system for organizing users into groups called enclaves rather than relying on topology or system-focused groups. Enclaves allow the administrator to organize networks based upon business functions to protect and grant access to critical servers and resources. FireMarshal application allows it to enforce policies for FireBlock and SmokeDetector on a very granular level from the entire network domain down to a single machine. The versatility of FireMarshal is seen in its ability to perform a complete security function from implementation to monitoring and reporting.

Finally, with security becoming a regulatory issue, an opportunity has opened up for Palisade to address the growing needs of this market. For example, Congress passed HIPAA in 1996 to standardize the electronic transmission of data within the field of health care. The gist of the law was that any organization in the private or public sector that handled medical records must comply with rules for electronic data interchange, privacy and security. There are as many as two million companies that will need to comply with the bill (Woo, 2001).

Conclusion

This field is destined to see rapid change, and it behooves technology managers to keep appraised of the laws and available products. This chapter highlights how one particular small player in the Internet security field has responded to recent important legislation requiring more stringent Internet and network security practices. Small players can offer an impressive array of products, and they should be included in a security product decision choice.

References

Mahanti, S., Bajwa, P., Strader, T. J., & Shrader, C. B. (2004). Palisade Systems: New markets for Internet security products. *Annals of Cases on Information Technology*, VI.

McGee, M. K. (2003). Rushing to get HIPAA extensions. *InformationWeek*, October 11, 2002. Accessed April 30, 2003 from: http://www.informationweek.com/story/showArticle.jhtml?articleID=6503824.

Palisade Systems. Accessed April 30, 2003, from: http://www.palisadesys.com/.

Symantec. (2002). Symantec Public Policy Update—March 2002—An Update on Policy Issues Affecting Symantec. Accessed April 30, 2003, from: http://enterprisesecurity.symantec.com/PDF/policy_march_02.pdf?EID=0.

Webdesk. (2002, March 18). Market Expected to Double and Remain Strong. Accessed April 30, 2003, from: http://www.webdesk.com/internet-security-software/.

Woo, S. (2001). Palisade helps park unwanted use of Net. *The Des Moines Register*, (April 26).

** Daniel M. Norris, CPA, MCSE, MCDBA, is a Visiting Associate Professor in Information Systems and Accounting at Drake University in Des Moines, Iowa. He holds a B.S. in Computer Science from Iowa State University, an M.S. in Accounting from Drake University and a Ph.D. in Accountancy from the University of Missouri-Columbia. He has work experience in computer programming and public accounting, and has taught courses in computer security, computer auditing and e-commerce security. Dr. Norris has teaching and volunteer experiences in Russia and in numerous Asian and African countries.*

** Philip A. Houle is Associate Professor of Information Systems in the College of Business & Public Administration at Drake University. He received his Ph.D. in computer and information control sciences from the University of Minnesota. Professor Houle teaches Database Management, Data Communications and Networking, and Introductory Information Systems. He also served as the top information technology administrator at Drake University from 1999-2001, a time during which the university upgraded its campus network to improve campus security and network performance. Currently he is working on issues involving e-mail address identity/mobility and e-commerce, including the use of spy-ware.*

** Charles B. Shrader earned a Ph.D. in 1984 from Indiana University. He is currently Professor of Management at Iowa State University. His research program focuses on relationships of strategic management variables, such as formal planning with organizational performance measures. His research appears in the* Journal of Management, Human Relations, Entrepreneurship: Theory & Practice, Journal of Business Ethics, Corporate Governance, Journal of Business Research, Journal of Small Business Management, Case Research Journal, Decision Sciences Journal of Innovative Education, *and* Annals of Cases on Information Technology. *He has received numerous teaching and case writing awards including the 2002 Philip G. Hubbard Education in Iowa Outstanding Educator Award.*

Chapter IV

Personal Information Privacy and EC: A Security Conundrum?

Edward J. Szewczak, PhD
Wehle School of Business
Canisius College, USA

Biography

Edward J. Szewczak is Professor of Information Systems at Canisius College in Buffalo, NY, USA. His published research has appeared in the Information Resources Management Journal, *the* Journal of MIS, Information & Management, Data Base *and the* Journal of Management Systems, *among others. He has co-edited five scholarly books of readings for Idea Group Publishing, including the recently released* Managing the Human Side of Information Technology *and* Human Factors in Information Systems. *These two books contain more extensive treatments of the Internet privacy issue and its importance to e-commerce security.*

Executive Summary

The issue of personal information privacy (PIP) and e-commerce (EC) continues to be debated within the community of Internet users. The concerns of privacy advocates conflict with the concerns of technology growth advocates. The challenges to PIP posed by various forms of EC technology are not the result of the technology itself. Rather it is the uses of the technology that pose the threat to the integrity of PIP. In particular, the surreptitious monitoring of user behavior without the user's consent and the possible misuse of the collected information are the biggest threats to the growth of EC.

Introduction

PIP is arguably the most important issue facing the growth and prosperity of EC. A director of IBM's Global Trust and EC services unit has been quoted as saying that privacy and security are the largest inhibitors of progress for EC today.

In his excellent study on *Privacy in the Information Age*, F.H. Cate adopted the definition of privacy as "the claim of individuals, groups or institutions to determine for themselves when, how and to what extent information about them is communicated to others." This definition is interesting because it allows for flexibility in discussing privacy within the context of the Internet. Whereas many people worry about divulging personal information electronically, other people seem more than willing to give it away, trading their personal information for personal benefits such as free shipping and coupons.

Protecting PIP has ignited a debate that pits privacy advocates against technology growth enthusiasts. The results of a 1998 survey conducted by

Louis Harris & Associates, Inc. revealed that worries about protecting personal information ranked as the top reason people generally are avoiding the Web. A 2000 telephone survey conducted by Harris Interactive found that 57% of Internet users favor laws regulating how personal information is collected and used by Internet companies. A study by Cheskin Research and Studio Archetype/Sapient found among other things that Internet users in the United States, Latin America and Brazil perceived threats to their personal information integrity and money from predatory individuals as well as predatory institutions. A survey by NFO Interactive found that the safekeeping of online consumer personal information was the main reason people chose not to shop online. A survey by Jupiter Communications found that roughly 64% of respondents do not trust a Web site even if it has posted a privacy policy. The main concern was the handling of credit card data.

"Carders" buy and sell credit card numbers stolen from the Internet using Internet chat rooms. The carders announce a list of cards with accompanying personal information including billing address and phone number. The credit card numbers with accompanying information are usually purchased in short order. Possible uses of these numbers includes identity theft.

Failed Internet companies such as Boo.com, Toysmart.com and CraftShop.com have either sold or have tried to sell customer data that may include phone numbers, credit card numbers, home address and statistics on shopping habits, even though they had previously met Internet privacy monitor Truste's criteria for safeguarding customer information privacy. The rationale for the selling was to appease creditors. The defunct political portal Voter.com announced intentions to sell 170,000 e-mail addresses together with party affiliations and issues of interest.

On September 9, 1999, *Privacy Times* published the equation "Good Privacy = Good E-Commerce (& Vice Versa)." As events continued to unfold, it became increasingly clear that privacy concerns were plaguing EC. Wall Street began to revalue Internet companies that accumulated customer personal information to target marketing efforts. The FTC told a Senate panel that there were more than 300 online privacy bills–to limit the collection and "mining" of personal data–pending before state legislatures and Capitol Hill. *Business Week* called the privacy backlash "the privacy penalty." Consumer reaction included an unwillingness to click on Web site banner ads, which in turn led to advertisers becoming dissatisfied with Web portal effectiveness.

PIP and EC Technologies

Government regulators and enforcement officials have to consider a host of technological challenges to PIP on the Internet. These include technologies used in corporate and government databases, e-mail, wireless communications, clickstream tracking, CPU design and various biometric devices.

Corporate and Government Databases

The practice of gathering personal information about customers and citizens by corporations and governments is well established. Software is available which is dedicated to analyzing data collected by company Web sites, direct-mail operations, customer service, retail stores and field sales. Web analysis and marketing software enables Web companies to take data about customers stored in large databases and offer these customers merchandise based on past buying behavior, either actual or inferred. It also enables targeted marketing to individuals using e-mail. Governments routinely collect personal information from official records of births, deaths, marriages, divorces, property sales, business licenses, legal proceedings and driving records. Many of the databases containing this information are going online.

Financial Information Databases

The recent deregulation of the financial services industry has made it possible for banks, insurance companies and investment companies to begin working together to offer various financial products to consumers. Personal financial information that was kept separate before deregulation can now be aggregated. In fact the ability to mine customer data is one of the driving forces behind the creation of large financial conglomerates. Services can be offered to customers based on their information profiles. Banks that finance company alliances can disseminate personal information about their customers to third parties without their permission and may even decline to alert a customer if someone is snooping in the customer's account. Even though companies may choose not to sell personal information to third parties, companies within an alliance may use the data themselves to push financial products and services.

Large credit bureaus such as Equifax and Trans Union have traditionally been a source of information about a person's credit worthiness. Their databases contain information such as a person's age, address and occupation. Credit bureaus have begun to sell personal information to retailers and other businesses.

Online banking presents another challenge to PIP. Many banks store customers' addresses and social security numbers in the same records. The information, once retrieved, can be used to reroute credit card mailings or open new accounts.

Medical Information Databases

Like personal financial information, medical information is for most people a very private matter. Despite this fact, there is a wealth of personal medical data in government and institutional databases. Although many of these government database records are stripped of information which could be used to identify individuals (such as Social Security numbers), it is still possible to link the records to private sector medical records using standard codes for diagnoses and procedures employed by the United States healthcare system. The codes are usually included on insurance claims and hospital discharge records.

Much personal health information that is available to the public is volunteered by individuals themselves, by responding to 800 numbers, coupon offers, rebate offers and Web site registration. The information is included in commercial databases like Behavior-Bank sponsored by Experian, one of the world's largest direct-mail database companies. This information is sold to clients interested in categories of health problems, such as bladder control or high cholesterol. Drug companies are also interested in the commercial databases.

Medical information databases are available through private networks. However, this situation is quickly changing. Healtheon and other healthcare companies are competing to get doctors to write prescriptions over the Internet and to persuade people to place their personal health records on the Internet.

E-Mail

E-mail accounts for 70% of all network traffic and is susceptible to tampering and snooping. In many companies, employee e-mail communications are routinely monitored. Loss of workday productivity is often cited as the major concern for businesses that monitor e-mail. However, many companies worry about possible litigation stemming from sexually charged e-mail. Companies are also concerned with activity which may expose the company to breach of contract, trade secret, and defamation lawsuits.

Employee's invasion of privacy claims have not been upheld in the United States courts, which argue that, since employers own the computer equipment, they can do whatever they want with it. The 1986 Electronic Communication Privacy Act grants employers the right to review stored communications on a company's computer system.

Wireless Communications

A monitoring operation run by the U.S. National Security Agency called Echelon uses satellite technology to listen in on virtually all international and (to a limited degree) local wireless communications, including phone calls, faxes, telexes, e-mail and all radio signals including shortwave, airline and maritime frequencies. The operation listens for certain target words. When a target word is encountered, the transmission is sent to humans for analysis. Echelon is designed primarily for non-military targets, including governments, organizations and businesses around the globe.

Wireless advertising promises to pose a host of challenges for privacy advocates. Wireless service providers know customers' names, cell phone numbers, home and/or office addresses, and the location from where a customer is calling as well as the number a customer is calling. Each wireless phone has a unique identifier that can be used to record where in the physical world someone travels while using the cell phone. In addition, the Federal Communications Commission requires cell phone service providers to be able to identify the location of a caller who dials 911, the emergency number. Most likely cell phone manufacturers will meet this requirement by embedding a Global Positioning System chip in all cell phones. Since a cell phone service provider can track the location of a 911 call, it will also be able to track the location of any other call as well.

Clickstream Tracking

As with e-mail technology, productivity and legal liability concerns are also paramount in companies' decisions to track the behavior of employees when using the Internet. Software programs have been specifically designed to monitor when employees use the Internet and which sites they visit. Telemate.Net can examine company network activity and produce reports identifying and ranking the company's heaviest individual Internet users. It lists the sites most visited by members of the whole company or by members of individual departments within the company, and if desired can list sites visited by individual employees and rank them by roughly two dozen categories.

Internet companies monitor Internet user behavior by a number of means, primarily to gather data about shopping and buying preferences with a view toward developing "user profiles." These technological means primarily involve the creation and use of cookies. Cookies are text files created by a Web server and stored on a user's hard disk. A cookie is a set of fields that a user's computer and a server exchange during a transaction. Web servers work with ad placement companies that resell advertising space from popular sites. These companies maintain large databases in which are recorded details about who looks at which pages. When a user connects to a Web site, the browser checks the cookies on the hard drive. If a cookie matches the site's URL, the browser uploads the cookie to the Web site. With the information contained in the cookie, the site can run programs which personalize site offerings and/or track the user's activity while online.

It should be noted that U.S. government agencies also track the browsing and buying habits of Internet users. A congressional report released in April 2001 found that 64 federal Web sites used files that allow them to track the browsing and buying habits of Internet users. Among the agencies were the Departments of Education, Treasury, Energy, Interior and Transportation, as well as NASA and the General Services Administration.

Hardware and Software Watermarks

Hardware and software identifiers ("watermarks") can also be used to identify individual users. Every Ethernet card used in computer communications has its own MAC (Medium Access Control) address, a 48-bit number sent in the header of every message frame. As the Ethernet standard evolves into a wide-area communications protocol, this identifier may become of increasing concern to Internet users intent on protecting their privacy.

Microsoft Corporation includes a unique numeric identifier into every copy of its Office program. When a Microsoft Office document is created, it is watermarked with this unique identifier. The creator of the Melissa virus was apprehended when he posted documents to a Web site frequented by virus makers. Authorities used the watermark found in the Melissa virus to match the watermark found in the documents.

Biometric Devices

Various devices are available that identify people through scans of their faces, hands, fingers, eyes or voice recognition. Biometric devices create a statistical profile by assessing a number of biological characteristics. As the equipment used to take the measurements decreases in cost, it becomes economical to scan millions of faces and other characteristics into a computer database. Digital photography adds to the growing volume of non-text data about people. Privacy advocates object to the fact that much of the measurement taking happens without the knowledge or explicit cooperation of a subject, which can lead to abuses of the technology. The Electronic Frontier Foundation has noted that a bank that has collected face scans of ATM customers could sell this information to another company for a purpose not related to banking. Though not as simple as text data, biometric data can be transmitted on the Internet with little difficulty.

Conclusion

Privacy is a social issue, generally speaking. How the PIP debate is ultimately resolved will be decided by the values inherent in a society. Since the position of the privacy advocates differs so markedly from the position of the technology growth advocates, and since privacy issues have been addressed in court and precedents established in state and common law, it seems likely that the PIP debate will be resolved in the world's legislatures and laws produced there will be enforced in the courts. Assuming the privacy concerns of individuals do not bring EC to its knees, technology will continue to grow and be used to bolster the growth of EC. Any laws that are passed must take into account the evolving nature of technology, while at the same time respect the PIP values of individuals.

Recommended Readings

Cate, F.H. (1997). *Privacy in the Information Age*. Washington, DC: Brookings Institution Press.

Lessig, L. (1999). *Code and Other Laws of Cyberspace*. New York: Basic Books.

Chapter V

Developing Secure E-Commerce in China

Michelle Fong, PhD
School of Applied Economics
Victoria University, Australia

Biography

Dr. Michelle W. L. Fong is a lecturer in the School of Applied Economics, Victoria University. Prior to her academic and research career, she worked with different business systems in different corporations in Singapore, Malaysia, China and Australia. This gave her an insight into the information technology applications within these organizations, which spurred her research interest in the e-commerce field.

Executive Summary

The Chinese government has been keen to develop electronic commerce (e-commerce) as a source of economic growth and modernization. While B2B (business-to-business) online transactions are boosted by state-owned enterprises or government-affiliated businesses, B2C (business-to-consumer) online transactions constitute a very minor proportion of e-commerce activities. Several obstacles have deterred consumers from embracing the Internet for B2C online trading and payment, such as inconvenient electronic payment systems, low public confidence in the insecure electronic networks and inadequate regulatory frameworks. It is imperative for the progress of e-commerce in China that electronic systems are secure, and operating frameworks transparent and stable. Otherwise, this emerging market economy will lag behind in enjoying the economic benefits of e-commerce, an emerging key facet in the World Trade Organization (WTO) environment.

Introduction

Internet use in China has surged from 620,000 users in October 1997 to 59.1 million users in December 2002 (China Internet Network Information Center, 1997, 2003), giving the country the third largest number of Internet connections in the world, after the USA and Japan (*The Economist*, 2003). The 59.1 million Internet users represent 4.6% of market penetration in the Chinese economy, which indicates an enormous untapped market potential for e-commerce. Developing e-commerce as a source of economic growth and modernization has been of great interest to the Chinese government, who promulgated it at the national strategic level in the 2001-2005 Five-Year-Plan. In this Five-Year-Plan, e-commerce was envisaged to grow from US$9.32 billion in 2000 (Lawson, 2001) to more than US$20 billion in 2005 (Healy & Duke, 2002). Despite growing optimism, e-commerce is still in its infancy and the task of realising its economic potential remains a real challenge for China. In 2001, the Economic Intelligence Unit (2001) ranked China in 45th position in its survey of the 'e-readiness' of 60 countries and classified this emerging market economy into the category of 'e-business laggards'. It was suggested that countries in this category have to overcome significant hurdles in the development of e-commerce.

Discussion of Issues

In 2000, the total revenue generated by e-commerce transactions in China constituted 0.87% of GDP. B2B transactions accounted for 99.5% of this revenue, whereas B2C transactions accounted for 0.5% (www.china.org.cn, 2001). B2B transactions accounted for a substantial proportion of e-commerce revenue because they were mostly conducted by state-owned enterprises or affiliated businesses, which still dominate the current economic system and operate under the influence of the government's directive for e-commerce development. The low level of B2C activities was the result of consumers' reluctance or hesitation in embracing Internet as a channel of trade. The failure to reach critical mass consumer acceptance for B2C online trading has been largely attributed to the absence or slow improvement of a number of critical success factors, such as public confidence in the security of online systems, technological support and regulatory frameworks. In e-commerce, a secure and efficient electronic payment system is a vital support for online trading because prompt settlement contributes to the stability and liquidity of the economic system. In addition, information is a critical strategic resource in this electronic medium, and therefore a stable and transparent operating environment is essential for instilling sufficient confidence and trust among participants in the transmission of important or confidential data. An underdeveloped electronic system is an easy target to external hackers or fraudulent insiders, whose abuse can result in unauthorised access to confidential information, unauthorised transfer of funds, loss of important data or significant operational disruptions and losses.

In a survey conducted in December 2002, 0.1% of Chinese Internet users cited online shopping as the main reason behind their Internet subscriptions (China Internet Network Information Center, 2003). Although a higher proportion of Internet users made online purchases in 2002 (33.8%) than in 1999 (8.99%), the number of users who are inclined to conduct or repeat online purchases in both years has remained below the level of interest indicated in 1998 (China Internet Network Information Center, 2003, 2000). Online shoppers who have used electronic payment systems to settle their online transactions have voiced serious concerns about network security, inconvenient payment systems and the trustworthiness or authenticity of the Internet sellers.

Cyber Risk

The increase in the number of computers connected to the Internet in China has been accompanied by an increase in the number of illegal activities involving computers and the Internet. Computer crime activities were reported to have gone up by 30% annually in the late 1990s in China (*WorldNews*, 1999). It was estimated that the actual computer crime rate was six times higher than the number of reported or detected cases (Kabay, 2001). A survey found that 59% of Chinese Internet users experienced computer invasion from hackers in 2002 (China Internet Network Information Center, 2003). The vulnerability of computer networks to hackers has largely deterred consumers from embracing e-commerce as a channel of trade. In the financial system, the number of fraudulent activities and crimes committed over the computer network has been on the increase too. According to Xinhua News Agency (2003), the financial industry suffered losses of more than 10 billion yuan (US$1.2 billion) each year, as a result of virus attacks or hacker invasions. In a separate study, it was found that at least 90% of reported cases of computer network crime committed between 1997 and 2000 in this industry were due to inadequate security control in the system (Chen, 2002). Chinese consumers are very cautious of the insecurity involved in online bankcard payments through the Internet. It is estimated that the value of retail consumption transacted and settled through the Internet (online bankcard payment and online purchase) in 2002 was 0.15% of the total value of all retail consumption transacted through bankcards (online and off-line purchases). The scepticism of consumers about electronic bankcard payment systems for e-commerce can be understood from the level of bankcard use and acceptance in off-line purchases. Of the 2.7% of Chinese retailers contracted to accept bankcards issued by the different local banks in the country, 30% avoided bankcard as a payment instrument from customers, offering various reasons (Chui, 2002). In addition, 50% of the 460 million bankcards issued were dormant cards (Zhou & Wu, 2002). Clearly, many Chinese prefer the traditional way of paying in cash for their goods.

Weak Technological Support

The absence of an effective coordinating force and a coherent strategy, at both the organizational and national levels, and at the onset of the payment technology adoption process in the economy, has resulted in incompatible technologies and non-interoperable networks. Manual processing is often used to supplement the quasi-automated payment system, but the manual process of bridging the gaps between non-interoperable networks undermines the security and integrity of the bankcard payment process, making it harder to prevent the unauthorised use of bankcards, for instance.

In the event of card loss or cancellation, the prevention of unauthorised use is time-consuming and tedious, especially in areas where support infrastructure is inadequate. For example, when bank branches receive the weekly 'blacklist' (with details of invalid or suspended bankcards), they have to reproduce the same list for each of their contracted agents (who were contracted to accept bankcard for payment of transactions). This manual process takes up time and can create substantial risk in the electronic system if the agent does not receive the relevant information in time to stop the fraudulent use of such cards. In addition, the nascent legal framework can expose the bank, agent and cardholder to high financial risk. The unreliable electronic payment system and inadequate legal structure keep not only the cardholders from using the Internet as a channel of trade, but also the agents from relying on it for establishing online businesses.

Resource Constraints

Resource constraints and underdeveloped security controls within local banks have hampered the development of an efficient and secure electronic payment system to support e-commerce. For example, the inadequate bankcard credit verification facility (to support timely credit checks for all purchases) has exposed the banks to high credit risks. Furthermore, the lack of a credit information system in China has hindered the development of the card business for e-commerce.

A majority of the bankcards issued in China are debit cards, through which a cardholder can close transactions out of a preexisting credit balance, held by the card-issuing bank, to cover a certain value of the total transaction. But because the electronic payment system has not been capable of real-time processing or performing online credit limit checks, overdrafts have been a serious concern to the local banks. The banks' credit collection and pursuit of repayment of overdrawn amounts have been made difficult by outdated or fictitious records furnished by defaulting cardholders when they applied for a bankcard. In addition, the inadequate effort on the part of the banks in verifying and updating personal records on a consistent basis has aggravated the situation. In such cases, these banks have reluctantly become the bearers of losses.

Because of competition, banks do not share information on the credit standing or background of their cardholders or card applicants. As a result, there is duplication of resources for the same investigation procedure undertaken by the different banks on the same bankcard applicant. The proprietary nature of information on bankcard applicants has led to cases where applicants had successfully applied for various bankcards from different banks, because the banks were not able to detect the numerous applications lodged or the fraudulent information submitted by the defrauding applicants.

Regulatory Framework

The lack of sound bankcard payment and e-commerce regulatory systems has further slowed the progress of e-commerce. Although the Chinese government has been reviewing and tightening its legislative structure in support of the development of e-commerce, substantial effort is still needed to inject stability and confidence into the operating environment. For example, electronic contracts are recognised under the law of contract in China, but there is no reliable legislative framework for the use of digital signatures in online contracts. In addition, the banks have found that the existing legislation does not help them to recoup their losses without incurring high legal costs when bankcard guarantors shirked their responsibilities or agents breached their fiduciary position with the banks. The inadequacies in both legal and technological structures have caused banks to adopt a risk-averse attitude in the development of their bankcard businesses. The local banks tend to be overprotective of their own interests, compared to international card issuers, to the extent that their customers have to bear a higher burden of risk in the event of bankcard theft and unauthorised use of the stolen card. In order to brace themselves against economic loss or liability, most banks include a condition in the bankcard agreement with the cardholders that the latter are liable for any unauthorised payments made before, and even a day after, they notify the issuing bank of the loss of their cards (Wang, 2002). In addition, cardholders are required to pay a 40-yuan (US$4.80) processing fee when lodging the notification of card loss at the bank.

Implications

An efficient and secure electronic payment system and a stable and transparent operating environment will help China to play an important domestic and global role, as a formal member of the World Trade Organization (WTO). As a member, China is obligated to gradually open its sectors to overseas investors and implement market-based mechanisms for free and fair competition. The opening of markets for foreign participation entails the need to create stability and predictability in the operating environment so that investors can plan their activity in the Chinese market with greater certainty. Secure electronic systems and robust operating legal structures will help to maintain market confidence for the sound functioning of both financial and economic systems within the emerging market economy. On the global front, e-commerce is becoming a key facet in globalisation, and robust electronic payment mechanisms will help China to integrate with this increasingly electronic-based and competitive global trading system. It is imperative that the local banks cooperate to resolve the incompatible standards and non-interoperable payment systems, because non-traditional competitors (foreign companies) with sophisticated networks and innovative product offerings are increasingly entering the domestic markets in this WTO environment. Otherwise, the electronic payment system and the economic system will not be able to cope with the pressure generated by changes in market structure and business practices in a fully open domestic market or a competitive global market.

Alternative Solutions and Recommendations

A comprehensive and integrated information technology support system is of critical importance for a flexible, convenient, secure and fast online payment system. Although there are several alternatives for overcoming network infrastructure inadequacy, they all have limitations. For example, the mobile phone network is an emerging technology that has the potential to deliver payment convenience, allowing banking access anywhere and anytime. Although mobile banking services are being offered by most of the major local banks, they are still in their early stages of development. In addition, security issues remain a major concern. A promising new technology is the smart card technology, which offers advantages over magnetic strip card technology, especially in terms of its increased security and wider application potential. However, the magnetic-strip bankcard constitutes at least 75% of all the bankcards issued in China. To replace the magnetic-strip card with the smart card requires careful analysis in terms of the cost-benefit equation. The replacement of the relatively cheaper magnetic-strip technology with smart card technology means more capital investment, with an uncertain or prolonged payment period for the banks. Although the smart card is increasingly being introduced in greater numbers and more cities, its operating framework is not as established as the magnetic-strip card. Unless there are means of financing a rapid transition to the smart card, it is expected that the magnetic-strip bankcard will remain the dominant card for some time.

A strong and effective coordinating force is needed to consolidate and connect the disparate networks. In March 2002, the government affirmed its desire to achieve a national and universal electronic payment system by 2005 through the establishing of China UnionPay Company Limited. This company was formed through shareholding stakes from the four major local state-owned commercial banks, 10 shareholding banks, 45 city commercial banks and 12 rural credit unions; and the structure was envisaged to provide a good platform for fostering cooperation in network consolidation and interconnection. The network integration effort took a further step in June 2002 when this Chinese entity became a member of Visa International and Mastercard International. This allows local members of the China UnionPay Company Limited to access and utilise the global bankcard processing networks of the two experienced international card issuers, for payment settlement and processing of international bankcards. It also provides an opportunity for the local members to leverage on this exposure to improve their domestic bankcard payment systems for e-commerce development.

Credit reporting agencies and bureaus play an important role in gathering information for the local banks to make accurate and timely credit decisions. The existence of such entities would help alleviate the resource constraints faced by the banks in investigating the credit standing and authenticity of the information lodged by bankcard applicants. In addition, these agencies can help to eliminate duplication of the resources expended by each bank on the same applicant. However, such entities are new to the emerging market economy and may require the participation of established and experienced foreign companies. The injection of foreign expertise may help the bankcard industry to adopt a consistent framework for assessing the credit standing and risk of the applicants, which is presently highly inconsistent among the local banks and below the standard of international players.

The Chinese government should continue to play an active role in creating legal and institutional frameworks that will respond promptly and appropriately to the changes brought about by development in technology and e-commerce. In particular, there is an urgent need for frameworks that address the duties and rights of the participants, and the verification and certification of electronic signatures for online contracts, in order to cultivate widespread acceptance of online payment and trading.

Conclusion

While the Chinese government is enthusiastic about the potential of the Internet for e-commerce, the impressive Internet growth in China has not been accompanied by notable growth in B2C online trade. Low bankcard acceptance among consumers and relatively underdeveloped electronic payment systems have hampered the development of pure Internet-based online transactions. The aforementioned impediments are the result of non-interoperable banking networks, inadequate online security control and a weak e-commerce regulatory system, which together have rendered Chinese banks unable to support online transactions and Chinese consumers unable to trust the electronic network for online trading. China has been showing strong signs of becoming one of the top economic powers in the world and cannot afford to be left behind when e-commerce becomes a key element in globalisation. The major ramification for the Chinese government is that if the country were unable to provide a stable and transparent infrastructure and operating environment, as well as secure payment systems and protocols in a fully open and competitive market, the idea of using e-commerce as a source of economic development would be meaningless.

References

Chen, J. (2002). Speed up the development of informatization is the responsibility of the central bank. *Financial Computer of China.* Retrieved April 19, 2003, from: http://www.fcc.com.cn/computer/j200212/j021204.htm.

China Internet Network Information Center. (1997). Statistical report of the development of China Internet (1997.10). *China Internet Network Information Center.* Retrieved April 14, 2003, from: http://www.cnnic.net.cn/develst/9710/e-9710.shtml.

China Internet Network Information Center. (2000). Semi-annual survey report on Internet development in China. *China Internet Network Information Center.* Retrieved April 14, 2003, from: http://www.cnnic.net.cn/develst/e-cnnic2000.shtml.

China Internet Network Information Center. (2003). Analysis report on the growth of the Internet in China. *China Internet Network Information Center.* Retrieved April 15, 2003, from: http://www.cnnic.net.cn/develst/2003-1/1.shtml.

Chui, J. F. (2002). Developing a reasonable structure of interbank fee and returns repatriation. *China Credit Card.* Retrieved April 9, 2003, from: http://www.fcc.com.cn/card/k200212/k021208.htm.

The Economist. (2003). Survey: Caught in the Net. *47*(1), 16-17.

Economist Intelligence Unit. (2001). The Economist Intelligence Unit/Pyramid Research e-readiness rankings. *Ebusinessforum.com.* Retrieved April 30, 2003, from: http:// http://www.ebusinessforum.com/index.asp?layout=rich_story&doc_id=367.

Healy, T. R. & Duke, K. E. (2002). Winners and losers under China's e-sign law. *The China Business Review, 29*(6), 32-35.

Kabay, M. E. (2001). Studies and surveys of computer crime. Retrieved April 24, 2003, from: http://www2.norwich.edu/mkabay/methodology/crime_studies.htm.

Lawson, S. (2001, February 27). China e-commerce cools down. Retrieved May 23, 2002, from: http://ad.doub...=rectangle ;sz=336x280;tile=1;ord=330626?.

Wang, H. (2002). The main legislative problem concerning the use of bankcard and solution strategies. *China Credit Card.* Retrieved April 9, 2003, from: http://www.fcc.com.cn/card/k200212/ k021044.htm.

World News. (1999). China grapples with the Internet. Retrieved April 16, 2003 from: http://www.finalcall.com/international/china-internet1-26-99.html.

www.china.org.cn. (2001). E-commerce to better serve traditional industries. Retrieved April 16, 2003, from: http://service.china.org.cn/ link/wcm/Show_Text?info_id=11546&p_qry=B2C.

Xinhua News Agency. (2003, April 2). China seeks to build boundary on Internet. *www.china.org.cn.* Retrieved April 16, 2003, from: http:/ /service.china.org.cn/link/wcm/Sho...t?info_ id=60670&p_qry=computer and fraud.

Zhou, G. Y. & Wu, S. S. (2002). Several hurdles in the development of bankcard in China. *China Credit Card.* Retrieved April 9, 2003, from: http://www.fcc.com.cn/card/k200212/k021221.htm.

Recommended Readings

Fong, M. W. L. (2003). Bankcard payment system in the People's Republic of China. *Annals of Cases on Information Technology, 5*. Hershey, PA: Idea Group Publishing.

Chapter VI

Identifying and Managing New Forms of Commerce Risk and Security

Dieter Fink, PhD
School of MIS
Edith Cowan University
Australia

Biography

Dr. Dieter Fink is an Associate Professor in the School of Management Information Systems at Edith Cowan University in Western Australia. Prior to joining academe, he worked as a Systems Engineer for IBM and as Manager, IS Consulting for Arthur Young (now Ernst & Young). He has been a visiting academic at various universities including the Australian National University, Canterbury University and Free University of Berlin. His interests are primarily in establishing and measuring the value of IT investments, and minimising the risk of e-commerce and e-business applications.

Executive Summary

Under the system of e-commerce, organisations leave themselves open to attack, which can have catastrophic consequences. This is because new risks and insecurities have emerged brought about by changes in information technology and systems. This chapter outlines the characteristics of openness, virtuality and volatility associated with e-commerce systems and the technological security requirements, such as firewall software, encryption, and digital signatures and certificates. Implications are drawn for effective risk and security management (RSM). These include stakeholder participation, a holistic approach and gaining competitive advantages. From those, recommendations are made to management focusing on the effectiveness of the RSM approach, and ways to provide competitive advantages and a smooth implementation of RSM.

Introduction

Commerce in the New Economy is generally referred to as e-commerce, where organisations sell their products and services online to consumers, businesses combine with other businesses to form virtual enterprises and suppliers link with partners in a supply chain. The driving force behind the New Economy is increased digitisation of data and information enabled by the Internet and associated technologies, such as the World Wide Web (Web). The Internet is as a non-hierarchical, democratically structured, collaborative arrangement entered into by millions of network users representing consumers, businesses and government.

Organisations practising e-commerce potentially have much to gain, but leave themselves open to attack and compromise which can have catastrophic consequences. New forms of crime are developing (e.g., denial of service) which have not been experienced before. These inherent insecurities require that stringent risk and security management practices be adopted. The consequences of someone meddling with the website can range from mild (e.g., the introduction of a detectable virus) to catastrophic (a prolonged system outage leading to the loss of customers and eventual bankruptcy).

New Forms of Risk and Security

With the introduction of e-commerce, the Information Technology and Systems (ITS) environment has changed substantially and business is no longer conducted 'as usual'. While some of the risks associated with e-commerce are not new (e.g., hacking, theft of intellectual property), new insecurities have arisen because of the far-reaching scope of e-commerce. To understand the new risk environment, it is necessary to contrast it with that of the previous ITS environment.

- *Closed vs. Open Systems:* With previous generations of IT, systems were less accessible and open to attack. For example, damages to stand-alone systems and local area networks (LANs) are restricted in-house. E-commerce systems, on the other hand, provide increasing levels of connectivity and accessibility to data and networks from outside the organisation.
- *Tangible vs. Virtual Assets:* Traditional ITS environments are more tangible and were easily recognised as data processing centres. With e-commerce, information and virtual trading communities are more difficult to track. Intangible assets have become more important and take the form of intellectual property, information and knowledge.

- *Development vs. Operations:* Systems in the past were developed in a controlled manner and released for operations after extensive testing. With e-commerce, the need for market responsiveness requires that systems are developed and operated in a very short time. Operations have become critical because e-commerce aims at high transaction rates in order to bring down the costs of transaction processing.
- *Predictability vs. Volatility:* In the past, risk and security management could take place at a leisurely pace and reviews were conducted every couple of years. The RSM culture for traditional ITSs is unlikely to be satisfactory for the e-commerce environment. With each development of an e-commerce function, new elements of risk emerge and uncertainty arises.

Compared to the RSM processes of older ITSs, those for e-commerce have become more complex and greater interdependencies have to be considered. Furthermore, the nature of assets to be protected

Figure 1: The Processes of E-Commerce Risk and Security Management

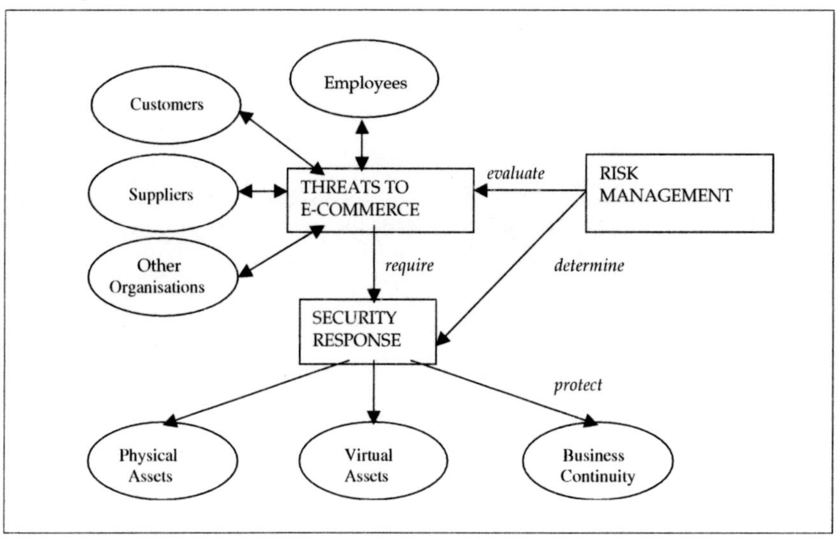

Table 1: E-Commerce Security Issues and Responses

Security Domain	Traditional Approach	E-Commerce
Access	Locks and keys, fences and walls	Firewall software
Confidentiality	Limit physical access to documents	Encryption
Authentication	Letterheads, written signatures	Identification and passwords Digital signatures and certificates
Integrity	Clerical checking and managerial control	Organisation controls Application controls
Attack	Theft of goods	Computer viruses Computer crime
Continuity	Manual processing and recovery	Electronic backup and recovery

has changed and business continuity has become critical. The changes are reflected in Figure 1.

Security responses to the e-commerce risks identified here have also changed, especially when compared to traditional approaches. They are reflected in Table 1.

As can be seen from the above table, a number of new technology-based security approaches are needed for e-commerce. They include firewall software which has the purpose of securing the internal "trusted" network from the external "untrusted" network through a highly monitored access point. The software provides essential protection against computer hackers. Other important technologies are encryption, where confidential and sensitive information is changed to protect content, and digital signatures and certificates which take the place of handwritten signatures and physical evidence of a person's credentials.

Implications for Risk and Security Management

For effective RSM the following key approaches should be adopted because of associated significant benefits.

Stakeholders Participation

RSM is often seen solely as the domain of IT professionals with security being viewed from a technological aspects, i.e., implementing technological measures such as firewalls and encryption. Kelly (1999) refers to this practice as "point solutions"–quick fixes that can do more harm than good. Management and users appear rarely involved in the process. With the increasing integration of all business activities, as occurs with e-commerce, stakeholders from all areas of the organisation need to be involved. In this way a balanced set of risk and security measures will be adopted that includes technology, people and procedures.

Holistic

"Many security problems are caused by all too human misperceptions of where dangers actually lie and the ability of particular measures to avoid them" (Brewer, 1999). The levels of security understanding of security threats, exposures, safeguards, practices and priorities among information users and solution providers varies widely. RSM therefore needs to be approached in a systematic manner so that all perceptions are included and evaluated. By capturing a wide variety of opinions, all facets of RSM will be implemented.

Competitive Advantage

RSM activities are seen as burdensome practices that create additional work for already stretched resources. They are often perceived to be only needed when the organisation is under attack or special circumstances arise. This negative perception needs to be reversed and RSM should be seen as an enabler since, by operating safely, the organisation can take more risks than its competitors. Furthermore, RSM has the obvious advantage of preventing expensive system outages, thereby ensuring the continued viability of the organisation.

Recommendations

A RSM approach to e-commerce should build on the fundamentals of good management principles. The approach can be summarised as follows.

Is the approach effective for e-commerce?

- *Comprehensive:* It must cover technical and business scenarios that are part of the various types of e-commerce (business-to-business, business-to-consumer), the phases of the e-commerce development (from planning to implementation) and the life cycle of operations (from ordering to supply and payments).
- *Inclusive:* The approach must cover all assets, vulnerabilities and threats. They include technology and business assets, real and virtual, by themselves as well as their interactions.
- *Flexible:* It must offer a variety of techniques that can be applied across some or all phases of e-commerce. Traditionally, an organisation's assets and policies may have been static, but threats in e-commerce are mobile and mutable.
- *Pro-Active:* The methodology must be flexible and promote pro-activity to anticipate changes in the e-commerce environment. It should encourage pro-active behaviour that uses RSM to gain competitive advantages.
- *Relevant:* RSM should lead to the identification and application of security measures relevant to e-commerce. Security techniques for e-commerce include the installation of firewalls, the use of digital signatures and certificates, encryption, etc.

Will the approach provide a competitive advantage?

- *Value:* The cost of RSM should be covered by the benefits realised from its use. In the real world, resources are limited and decisions about trade-offs have to be constantly made. RSM should be justified in terms of the advantages it provides.
- *Integration:* With e-commerce it is imperative that decisions are made based on both business and technological considerations. Risks in the technological domain interact with those in the business domain, and RSM should cover both types of risk.

Can the approach be implemented readily?

- *Systematic:* The processes of RSM should be structured and systematic to encourage organic management behaviour, transparency and open communications. Guidelines should be available for processes to be followed and the deliverables to be produced for each activity and phase.
- *Adaptable:* RSM must be integrated into the existing ITS environment, organisational culture and resource constraints with the objective of making an uncertain environment more certain.
- *Timely:* RSM must be carried out speedily because of the rapid changes that can occur for e-commerce. It must therefore define procedures, deliverables and timeliness that can be applied to small as well as major changes.
- *Tracking:* With increased operational risk emerged the need to measure and monitor risk factors through risk indicators and metrics. Effective RSM should provide the system for this and ideally produce dollar-at-risk-type figures.
- *Sponsorship:* It is generally accepted that projects fail if not supported by senior management. E-commerce RSM should therefore be an integral part of organisational risk and security solutions.

Conclusion

The chapter has shown that e-commerce differs significantly in its characteristics, risk profile and security requirements from those of traditional ITSs that have previously underpinned commerce. RSM has become more critical since with e-commerce, the need for market responsiveness requires that systems are developed and operated in a very short time and continue to operate. Should a system outage occur due to a security breakdown, the time to recover is very short as well, since customers will be able to choose from other suppliers on the Internet to meet their needs and probably will not return.

To ensure effective RSM in an e-commerce environment requires a new approach by management. They need to realise that the new characteristics of openness, virtuality and volatility mean that a broad set of stakeholders participate in developing risk and security approaches. Their involvement will lead to a holistic approach where all facets of RSM will be considered and implemented. RSM should not be viewed as a costly overhead since it can provide competitive advantages in the form enabling new opportunities offered by e-commerce to be realised with confidence and ahead of rivals.

In the future, the speed with which new e-commerce architectures and applications will emerge will increase, and the danger exists that these developments may well outstrip the ability of management to keep up with the task of securing them. If they can't, it will severely limit the organisation's ability to exploit new business possibilities, causing the organisation to fall behind its competitors. The recommendations made in this chapter are sufficiently broad to guide management for the foreseeable future in managing the risk and security requirements of their e-commerce systems.

Recommended Readings

Brewer, D. (1999). Keeping virtual worlds open for business. *Telecommunications, 33*(9), 65-66.

Davies, R. (1999). Taking the risk out of e-commerce. *Techwatch*, (October), 45-46.

Fink, D. (1997). *Information Technology Security—Managing Challenges and Creating Opportunities.* CCH Australia, Sydney.

Fink, D. (1998). *E-Commerce Security.* CCH Australia, Sydney.

Hays, D. (2000). Insurers cover hackers' threat to e-commerce. *National Underwriter,* 20-25.

Janczewski, L. (2000). *Internet and Intranet Security Management: Risks and Solutions.* Hershey, PA: Idea Group Publishing.

Kelly, B.J. (1999). Preserve, protect, and defend. *The Journal of Business Strategy, 20*(5), 22-25.

McEachern, C. (2000). Infinity Launches E-Commerce Strategy, Internet Risk Applications. *Wall Street & Technology*, (May), 62.

Salehnia, A. (2002). *Ethical Issues of Information Systems.* Hershey, PA: Idea Group Publishing.

Chapter VII

E-Commerce Security and the Law

Assafa Endeshaw, PhD
Nanyang Business School
Singapore

Biography

Dr. Assafa Endeshaw has Ph.D. and LLM degrees from London University, and his LLB degree from Ababa University. His work at London University centered on intellectual property policy and technology-related legal disciplines such as information technology and transfer of technology laws as well as in international trade and franchising. He was previously a legal attorney and advisor in government departments, and a researcher and consultant in a law firm in Ethiopia.

Dr. Assafa has written more than 35 major articles and presented upwards of 17 conference papers that have appeared in publications across Asia, Europe and America. He has authored Intellectual Property Policy for Non-Industrial Countries *(Dartmouth, 1996),* Intellectual Property in China: The Roots of the Problem of Enforcement *(Acumen, 1996), (jointly)* Marketing and Consumer Law in Singapore *(1999) and* Internet and E-Commerce Law: With a Focus on Asia-Pacific *(Prentice Hall, 2001). His areas of interest are: intellectual property, information technology (including regulation of the Internet and e-commerce), international trade law, franchising law, transfer of technology law and consumer protection law*

Presently, he is an Associate Professor of Law at the Nanyang Business School of the Nanyang Technological University, Singapore.

Executive Summary

The nature of the Internet as an open network means that it is devoid of central control and regulation. That in turn has exposed the Internet to the caprices and untoward intentions of some of its participants. Online businesses particularly suffer from an explosion of fraudulent activities and breach of security (destruction or theft of data and identity). The law has attempted to catch up with the problems by providing sanctions against perpetrators. Alternative solutions such as technical means and ethical codes of conduct are also in place. However, the creation of a secure network demands more than law or better technology. There is widespread recognition among lawyers, management and information system specialists that the creation of a secure network is part of the broader task of creating a security culture, starting from top management and sustained by clear and easy-to-implement policies.

Introduction

The use of the Internet for business has been growing in spite of the dot.com bubble a few years ago. The open nature of the Internet enables connections across networks and individual computer terminals virtually on a global scale. Unfortunately, the absence of central control and regulation leaves the Internet exposed to the caprices and untoward intentions of the participants. In particular, it has invited current forms of illegality and business malpractices to migrate to the new sphere.

The types of illegality that thrive on the Net have been growing. The most widespread form is fraud. Fraud is committed through Web auctions, sale of online merchandise (including books, computer equipment and software), the provision of all kinds of Internet-based services (free e-mail, credit cards, advice, prizes and sweepstakes, stock quotes) and business opportunities (franchises, work-at-home schemes, pyramid sales). Consumers are subjected to the rigging of prices and imposition of undisclosed charges, the non-delivery of purchased goods or services, supply of defective products and all kinds of empty promises for which payment will have been made in advance. Often the perpetrator would be in another country or jurisdiction and the consumer could not do anything about the fraud.

Breach of security of networks and computers (known under the general name of 'hacking') is another form of illegality. Perpetrators of such acts go beyond mere browsing through networks for fun and excitement; they engage in destruction or theft of data, with ulterior motives of financial gain or plain sabotage. Hacking thus becomes a means of stealing vital information or data belonging to corporations or physical persons with a view to furthering the wishes of the hacker. Various forms of hacking have emerged over the years. Among them are eavesdropping on networks including through the planting of programs in the target network and inserting viruses into computer systems ("bombs") to vandalise the contents.

Both forms of illegality (fraud and hacking) have mushroomed along with the growing popularity of the Internet. There are no authoritative statistics on the incidence of forms of illegality in cyberspace because of the absence of agencies empowered to undertake such a task across the globe. However, surveys of all kinds have emphasised the need to tackle them if the Internet is to continue to expand as an avenue for business and general social intercourse.

In the U.S., the National Consumer's League routinely receives and logs complaints under the various headings of illegality mentioned above. According to the *2002 Internet Fraud Statistics* compiled by the Internet Fraud Watch[1] (of the U.S. National Consumer's League), fraud committed on online auctions and general merchandise sales constitute by far the most widespread (90% and 5%, respectively) of the top ten forms. The remaining forms of fraud involve Nigerian money offers, sale of computer equipment or software, Internet access services, travel/vacations, work-at-home schemes, credit card offers and advance fee loans. The most frequent means of perpetrating these frauds were websites (94%) and e-mail (6%).

Reports of breach of security of networks and invasion of privacy abound. *The Internet Security Threat Report*, for one, estimated that "companies averaged 30 attacks per company per week over the past six months" to February 2003.[2] It pointed out that 78 percent of all cyber attacks were a blend of various forms: "viruses, worms, Trojan horses, and malicious code with server and Internet vulnerabilities to initiate, transmit and spread an attack."[3]

Web-based businesses in particular have been concerned about the problems of fraud and breaches of security. The latter more than the former has remained the most important concern for such businesses and tops all efforts at making the Internet safe for business.

Discussion of Issues

Security risk in networks may originate in computing failures such as the presence of bugs or lack of systemic integration that leaves gaps that intruders exploit to gain unauthorized entry. Once they break into systems, intruders may then plant programs that modify, destroy or spy on the system for further gains. The intruders might also steal data or information, or engage in network eavesdropping for the same ends.

However, it must be stated at the outset that the problem of security cannot be confined to the manner of use or abuse of a specific technology.

It is part of the generic question of maintaining and generating trust and confidence in the online business environment; this becomes more acute with changes in technology becoming more pronounced and pervasive. The Internet has led to a disruption of existing security structures and demonstrated the need for change in that environment. Moreover, the concern for a more secure infrastructure merges with the security of the entirety of business operations. The variety of transactions running on different application systems and security requirements needs to be integrated holistically.

Legislative Measures in General

It has become pretty well established that the options for tackling security, as well as other general issues for e-commerce, revolve around applying or modifying existing laws or creating new legal instruments. Many nations introduced laws to curb computer misuses before the Internet acquired widespread appeal. It therefore became necessary to make adjustments in those laws to catch up with new forms of abuse that the Internet facilitated.

The Organisation for Economic Cooperation and Development (OECD) has set out policies and guidelines that member states would be expected to follow.[4] Of the most important group of states, the European Union and the United States, in particular, have formulated policies and taken steps to prevent such abuses on the Internet as the provision of harmful and illegal content, piracy of intellectual property, breach of personal privacy and system or transactional security.

The European Commission embarked on combating "cybercrime" as part of the eEurope initiative it launched in December 1999. Various action plans and policy frameworks[5] have been adopted by the European Council to enhance network security and establish a uniform approach to cybercrime.

Whilst it has not viewed the introduction of a legal instrument as a pressing need, it recognises the necessity, in the long run, of a harmonised approach towards the legal treatment of cybercrime (especially hacking and denial-of-service attacks) by member states. In the meantime, it encourages increased training for special law enforcement units. Recently, the European Commission has proposed the creation of a "European Network and Information Security Agency" to provide advice for member states in matters of cybersecurity.[6]

An interesting development is the Council of Europe's Cybercrime Convention (which has already been adopted). It provides for four categories of criminal offences: offences against the confidentiality, integrity and availability of computer data and systems; computer-related offences; content-related offences; offences related to infringements of copyright and related rights. Other than the consolidation of the various offences in one piece and creating uniformity across signatory states, the Convention offers little more than what is found in current laws of most nations. A positive development is that the U.S. has endorsed the main thrusts of the Convention.

In the U.S., the Computer Fraud and Abuse Act addresses unauthorized access, dissemination of malicious software and trafficking in stolen access devices (such as passwords). The law has been amended after the 9/11 events by the USA Patriot Act of October 26, 2001.[7] The amendment gives law enforcement agencies more leeway in mounting surveillance over, and more extensive powers of prosecution of, cybercrimes in furtherance of the protection of the national information infrastructure. It also raises the penalties in the previous Act.

More recently, the Institute for Information Infrastructure Protection (U.S.) has made recommendations in its report, "Cyber Security Research and Development Agenda," on identifying sources of attacks and securing those systems which have been affected by attacks.[8] The Bush Administration has also presented its "National Strategy to Secure Cyberspace," to enhance the protection of computer networks, in particular to prevent threats that could result in "debilitating disruption to our nation's critical infrastructures, economy or national security."[9] However, it has suggested no further regulation — a fact ceased upon by critics.[10]

Hacking

Unauthorised intrusion (hacking) into networks as well as private accounts has now become widespread. Very recently, hackers rummaged through the network of the University of Texas at Austin and accessed records of more than 55,000 students and staff.[11]

Police forces across the world routinely arrest and investigate hackers. Britain's National Hi-Tech Crime Unit in collaboration with its counterparts in the U.S.–the U.S. Secret Service and the FBI– arrested two men on charges of introducing worms into computer systems and causing damages around the world.[12]

The standard legal responses across nations have been to criminalise hacking. Three offences have been created: unauthorised access, access with intent to damage contents and unauthorised interception of services. A fourth offence was introduced by a small number of states (following the U.S.) to protect special networks (defence, finance and the like) from any attack.

Invasion of Privacy and Identity Theft

The breach of security of networks (including through interception) often involves the unauthorised appropriation of personal data and information. Such forms of invasion of privacy differ from other less invasive ways of collecting information (using surveys, deploying cookies), but their implications are similar. Intruders put such information to further illicit use by selling or disclosing it to others to cause harm to the subject of the information.

Identity theft is a form of hacking which results in possession of personal data and information by the hacker to masquerade as the true identity owner for further use. It has gained particular notoriety in recent years. The use of false identities to undertake activities in the name of the true identity owner has devastating consequences for the latter. Their credit cards are used for transactions to the benefit of the thief; their identities are used to apply for loans, mortgages and the like. Estimates of identity theft by the U.S. Federal Bureau of Investigation go as high as 500,000 in 2002. The U.S. Federal Trade Commission has registered 161,000 complaints of identity theft, that is 43 percent of all complaints.[13]

In many nations, fraudulent use of another's identity is generally dealt with under pre-existing laws sanctioning impersonation and misrepresentation. This may change with the mounting scale and breadth of identity theft taking place across nations. The U.S. introduced, in October 1998, the Identity Theft and Assumption Deterrence Act on a federal level, making it a criminal offence.

Piracy of Intellectual Property

The breach of network security may also involve unauthorised use of intellectual property (IP) belonging to others. The International Federation of the Phonographic Industry (IFPI) has reported a fall in global music sales of 7 percent (contrasted with 10 percent in the United States) which it ascribed to rising Internet piracy in 2002.[14] Recently, the British Phonographic Industry and IFPI have formally written to British universities complaining about the state of music piracy on campuses and the likelihood of legal action for lack of action to curtail it.[15]

The Motion Picture Association of America estimates yearly losses to piracy to be $3 billion.[16] The trial in Norway, on behalf of the Motion Picture Association of America, of a teenager who had written a computer program that allowed copying of DVDs came to nought earlier this year. The court apparently viewed the program to be legitimately used to copy movies that were legally purchased. The case has been appealed.[17]

The legal measures to combat online piracy of IP have been in the statute books of many nations since the introduction of new requirements through the 1996 Copyright Protocol of the World Intellectual Property Organisation (WIPO). The Protocol affirmed the need to protect online copyright and outlawed removal or tampering with rights management devices. Domain names have yet to be addressed in the same manner.

Alternative Solutions and Recommendations

The deployment of appropriate management approaches to maintain system security is critical. Undue reliance on information security technology might be counter-productive. The use of trusted third parties wherever necessary might also lessen the burden on businesses that might not have the resources to detect threats, respond to breaches quickly and effectively, as well as undertake vulnerability testing and security reviews.

A prominent role is nevertheless given to technological means of combating fraud such as devices that ascertain the identity of a credit card-holder. The use of encryption is one such method, though it is also open to fraudsters to use it to prevent detection. Biometric identification and electronic signatures also serve the same purposes. The U.S. argument that high-level encryption systems should not be publicly available or exported to other nations has not led to drought because other countries have moved in to fill the void. In any case, secure electronic payment systems require encryption. A by-product of this is the ability to control money-laundering operations.

Other technological mechanisms include firewalls, anti-virus programs and other software with specific capabilities such as preventing e-mail eavesdropping or piracy of intellectual property. Technical means have become increasingly prominent in the area of IP protection because the new technologies empower consumers to make copies at will or transmit them virtually to anyone. Microsoft has thus offered a "new Windows Rights Management Software (RMS)" to enable users to control "how and where different sorts of content" they generate could be further used. [18]

Moreover, national authorities have established computer emergency response teams to handle major attacks on networks and systems. These are ongoing activities and assume preventative roles. In the U.S., the FBI sought to introduce a system ("InfraGard") that alerts businesses about security threats to their information systems and provides them with safe e-mail communication to share concerns with each other. However, the fear of businesses suffering cyberattacks from being shut down for investigation by the FBI has limited its acceptance.[19]

A third possible alternative for securing networks is the rise of self-help schemes that establish standards of performance to which industry operators have to adhere. Various security, trust and privacy protection schemes have been established by groups of companies. Examples are TrustE and CPAWebTrust in the U.S., TrustUK and TRUSTEDSHOPS in Europe. These schemes hope to foster trust and confidence in networks and online businesses.

Conclusion

There is widespread recognition among lawyers, management and information system specialists that the creation of a secure network demands more than law or better technology. It is generally recommended that businesses create a security culture starting from top management and sustained by clear and easy-to-implement policies.

Regarding the role of national organisations that help protect networks, one problem stands out. The absence of a legal duty on the part of businesses to report attack diminishes the role of existing authorities (such as CERTs and the police) with the result that such important areas of business as finance and banking may remain more vulnerable than they would otherwise be. The need of accurate data for threat assessment must therefore be met if an improvement is sought in the current lacunae.

Endnotes

1 See http://www.fraud.org/2002intstats.htm.

2 Symantec Internet Security Threat Report Sees Sharp Increase in Reported Vulnerabilities but Drop in Overall Attack Activity. *Business Wire*, February 3, 2003.

3 *Ibid.*

4 The OECD Guidelines for the Security of Information Systems and Networks of July 25, 2002.

5 An example is the Commission's Communication: Creating a safer information society by improving the security of information infrastructures and combating computer-related crime. COM (2000) 890 final, January 26, 2001.

[6] [News] European cybersecurity agency planned. *ITWorld*, February 10, 2003. Available at: http://www.itworld.com/Sec/2199/030210eucybersecurity/.

[7] The official title is the Uniting and Strengthening America by Providing Appropriate Tools Required to Intercept and Obstruct Terrorism Act.

[8] Brock Read, Group Calls for More Academic Research in Computer Security. *Chronicle of Higher Education*, January 31, 2003. Available at: http://chronicle.com/free/2003/01/2003013101t.htm.

[9] Jonathan Krim, Cyber-Security Strategy Depends on Power of Suggestion. *Washington Post*, February 15, 2003, at E01.

[10] *Ibid.*

[11] Robert Lemos, Data thieves nab 55,000 student records. *ZDNet,* March 7, 2003. Available at: http://zdnet.com.com/2100-1105-991413.html.

[12] Two are held after computer virus raids. *Daily Mail* (UK), February 7, 2003. Accessed online from *Factiva*.

[13] David Ho, Government reports surge of identity theft complaints last year. Associated Press Newswires, January 22, 2003. Accessed online from *Factiva*.

[14] John Borland, Music industry: Piracy is choking sales. *CNET News.com*, April 9, 2003. Available at: http://zdnet.com.com/2100-1105-996205.html.

[15] Adam Sherwin, Universities to be sued over music downloads. *Times Online*, March 28, 2003. Available at: http://www.timesonline.co.uk/article/0,,2-625793,00.html.

[16] Norwegian teenager to face retrial for film piracy. *Reuters*. Oslo, February 28, 2003. Accessed online from *Factiva*.

[17] *Ibid.*

[18] Stacy Cowley and Paul Roberts, Microsoft makes documents more secure. *PCWorld*, February 24, 2003. Available at: http://www.idg.net/ic_1186054_9677_1-5046.html.

[19] David A. Vise, FBI takes aim at cyber-crime: Agency seeks to enlist wary private sector in joint prevention effort. *Washington Post*, January 6, 2001, FINAL A2. Accessed online from *Factiva*.

Recommended Readings

European Commission. (2001, January 26). Creating a safer information society by improving the security of information infrastructures and combating computer-related crime. COM (2000) 890 final.

Katyal, N. K. (2001, April). Criminal law in cyberspace. *University of Pennsylvania Law Review 1003*, 149.

The OECD Guidelines for the Security of Information Systems and Networks of July 25, 2002.

Chapter VIII

Rethinking E-Commerce Security in the Digital Economy: A Pragmatic and Strategic Perspective

Mahesh S. Raisinghani, PhD
Program Director, E-Business
University of Dallas, USA

Biography

Dr. Mahesh Raisinghani is a faculty member at the Graduate School of Management, University of Dallas, USA, where he teaches MBA courses in Information Systems and E-Business, and serves as Program Director of E-Business. Dr. Raisinghani earned his PhD from the University of Texas at Arlington and is a Certified E-Commerce Consultant (CEC). He was the recipient of the 1999 UD Presidential Award, 2001 King/Haggar Award for excellence in teaching, research and service. Dr. Raisinghani won the Organizational Service Award for being the Best Track Chair at IRMA 2002.

Executive Summary

The worldwide market for information security services will nearly triple to $21 billion by 2005, up from about $6.7 billion in 2000, according to International Data Corporation.

This trend that focuses on information security stresses the translation of strategic business objectives and models into an information systems architecture that combines data process, workflow, financial and simulation models. E-commerce security is a process, not an end result. Managers need to understand how advanced technology creates more robust, scalable and adaptable information systems for the organization dedicated to continuous improvement and innovation. The key questions for assessing status quo and modus operandi are discussed, and some alternative solutions and recommendations are proposed as we look ahead in the digital economy.

"If you're not changing faster than your environment, you are falling behind." –Jack Welsh, CEO of General Electric.

Introduction

In thinking about the broad theme of this book, i.e., to provide insight and practical knowledge obtained from industry leaders regarding the overall successful management of e-commerce practices and solutions, I would like to focus on the fundamentals that require a crystal clear e-commerce security strategy in a dynamic competitive environment of discontinuous change. This ensures that an organization's e-commerce security efforts and overall business goals and objectives are aligned. A clearly articulated and well-documented e-commerce security strategy can help organizations assess and evaluate the results of their efforts. The focus needs to shift from the needs-based assessment to the results-based assessment of e-commerce security.

The worldwide market for information security services will nearly triple to $21 billion by 2005, up from about $6.7 billion in 2000, according to International Data Corporation.

According to this study, the boom in the market will be driven by corporate desires for wireless access, extranets, and remote networks because new and greater security services will be needed to secure those technologies. E-commerce security is a process, not an end result. As the number of interconnections in the world increases, so too will the number of attacks. E-commerce security remains a relatively straightforward risk-management equation—the more security that is in place, traditionally, the more onerous it is for end-users.

Discussion of Issues

Prior research has demonstrated how someone who is capable of capturing less than an hour's network traffic could determine a user's private key. However, this is not the only potential problem with public key infrastructure (PKI) systems. Since any party may purchase PKI systems, a single party may purchase several units and send many encrypted messages back and forth to himself/herself. In this manner, such a party would have both the clear and ciphered text versions of the message. The missing third piece would be the encrypting key. However, in mathematics it is well established that if two elements of an equation are known, one can solve for the third. Here, dense computational tools may be required to crunch the sheer number of test results in order to determine how the keys are calculated, but the process is well founded and most likely known by most sophisticated governments and possibly others. Such a potential counter to a PKI system would put in jeopardy one of the most widely used tools for protecting information today. For this reason, governments closely control and guard devices used in their applications of public key type encrypted communications. Nevertheless, because of the overhead involved in managing and operating a private key system (key management, key generation, secure key distribution, key destruction, key storage and so forth), most individuals and businesses deem themselves incapable of efficiently and cost effectively implementing a private key infrastructure.

Other key issues with PKI systems are as follows:

- *Short Message Expansion:* some public key systems expand short messages. This provides a great deal of information to knowledgeable cryptanalysts.
- *Man in the Middle Attacks:* where A thinks he is negotiating a key with B, C stands in the middle and communicates with both A and B instead of A to B.
- *Non-Prime Number Keys:* some PKI tools have been shown to not generate prime number keys, thereby leaving them open to factoring attacks.

What are the effective solutions to these issues since, due to the overhead involved in managing and operating a private key system (key management, key generation, secure key distribution, key destruction, key storage, etc.), most individuals and businesses deem themselves incapable of efficiently and cost effectively implementing a private key infrastructure?

Limits on the strength of exportable encryption is necessitated by the dual requirements of governments to not only provide a means for their citizens' communications to be protected, but also to be able to read communications of those who would thwart the law and/or violate national tranquility. These juxtaposed requirements create the situation where governments must perform both offensive and defensive functions by balancing the requirements of each against the other. This conundrum has caused much concern for civil libertarians in the U.S. and elsewhere who claim their communications should be completely private. How do e-commerce security managers in governments abdicate their legitimate dual, bifurcated responsibilities?

How can industry and government work together to ensure that the highest security standards are developed and adopted? For instance, there will likely be lawsuits filed against U.S. public corporate entities by shareholders should such corporate entities suffer a loss due to a computer security breach under the premise that the Board of Directors and the officers of the corporation did not properly execute their fiduciary duties in using due and reasonable care in protecting corporate assets. Such potential litigation should be a catalyst for prudent management of U.S. public companies to become very proactive in overseeing the entity's information security plans and implementations. In fact, the possibility of such potential litigation may make using U.S. Government-approved CCTLs or other certified information security tools an imperative in order to show that the entity used tools certified by the U.S. Government – under the theory the government is the highest information security certification body in the USA.

From a meta-perspective, the key issue for managers to consider is the following:

If we were not already using this e-commerce security technology/technique, would we go into it today? OR Are we moving fast enough today to build our expertise in e-commerce security to exploit immediate opportunities for streamlining inter-company processes, outsourcing activities in which we do not have distinctive capabilities and/or designing e-commerce security systems that we can market to other companies?

Implications: Key Questions for Assessing Status Quo and Modus Operandi

The implications for management and their impact on business strategy can be best understood by the following key questions that one needs to ask and think ahead about the implications of e-commerce security for management:

1. Does our management team have a shared vision of the long-term business implications of the new e-commerce architecture in general and the new e-commerce security architecture in particular?
2. Do we have a transition plan that balances the state of the architecture's development with a clear understanding of the areas of highest business impact?
3. Do we have a clear understanding of the obstacles within our organization that may hinder us from exploiting the full value of the IT/ e-commerce architecture, and do we have initiatives under way to overcome these obstacles?
4. Are we exerting sufficient leadership in shaping both the functionality offered by the internal and/or external providers of e-commerce security and Web services (defining, for example, the performance levels required for mission-critical applications) and the standard needed to collaborate with our partners?
5. How will this status quo change when the technology arrives to make impenetrable security invisible to end-users?

Alternative Solutions and Recommendations: Looking Ahead in the Digital Economy

Although several standardization initiatives in the area of authentication have already been launched by standards bodies and industry forums at national, regional and international levels, it was ascertained that they lacked the necessary consistency and coherence for validity and cross-recognition. To remedy this, the European ICT Standards Board, with the support of the European Commission, has launched an initiative bringing together industry and public authorities, experts and other market players: the European Electronic Signature Standardization Initiative (EESSI).

EESSI seeks to identify under a common approach the needs for standardization activities in support of the Directive's requirements, and to monitor the implementation of the work program by ensuring that three main principles were adhered to:

- effective involvement of all parties concerned with the broad subject area of electronic signatures;
- openness and transparency of the mechanisms used and of the initiatives taken;
- encouragement of global, internationally accepted solutions while avoiding duplication of work

The Data Encryption Standard (DES) has a new replacement called the Advanced Encryption Standard (AES), which is theoretically more secure. Additionally, in the United States, the National Institutes of Science and Technology (NIST) (www.nist.gov) and the National Security Agency (NSA) (www.nsa.gov) have also worked with industry to establish five Certified Cryptographic Testing Laboratories (CCTLs) (http://www.nsa.gov/isso/bao/cpep.htm) that will test industry computer security solutions and issue U.S. Government (USG) specified security-level certifications.

New security techniques to protect the corporate network provide organizations additional layers of security (above and beyond firewalls and encryption), providing better overall security. This is especially true when they are optimized for a particular application, such as integrity of the Web servers, and treated as incremental solutions, not replacements to traditional network security measures. These innovative network security solutions include honey pots or decoys, air gaps, exit controls, self-healing tools and denial-of-service defenses.

Honey-pots are decoy services that can divert attacks from production systems and let security administrators study or understand what is happening on the network. For example, Mantrap, from Recourse, is an industrial-strength honey-pot deployed next to data servers to deflect internal attacks, and located off the firewall in the demilitarized zone (DMZ) to deflect external threats. Factors that impact its success are quality, naming scheme, placement and security policy.

The processes that an organization should have in place in order to ensure that transactions such as wire transfers, electronic investments, etc., proceed securely is to deploy honey-pots in quantities equal to or greater than that of the production system. Honey-pots can get expensive which is why companies must choose the critical servers they want to protect.

Air gap technology provides a physical gap between trusted and untrusted networks, creating an isolated path for moving files between an external server and a company's internal network and systems. Vendors include RVT Technologies, Spearhead Technology and Whale Communications.

Self-healing tools are security and vulnerability assessment tools that can detect and fix weaknesses in an organization's systems before problems occur. For example, Retina 3.0 from eEye scans the range of IP addresses provided by the network administrator for vulnerabilities, software flaws and policy problems, reports it and can repair the vulnerability locally or remotely.

Denial-of-service (DoS) attacks make computer systems inaccessible by exploiting software bugs or overloading servers or networks so that legitimate users can no longer access those resources. Vendors include

Arbor Networks, of Waltham, Massachusetts; Mazu Networks, of Cambridge, Massachusetts; and Asta Networks in Seattle, Washington. For instance, Mazu Networks' solution to distributed DoS attacks works via intelligent traffic analysis and filtering across the network. A packet sniffer or packet analyzer acts as a monitoring device to evaluate packets on the network at speeds up to 1 Gbit/second and determines which traffic needs to be filtered out.

Conclusion

Typical challenges posed by e-commerce security to businesses of all sizes, such as increased customer demand for more information, better service and more efficient transactions, will be put in context with e-business models. The effects of larger and more nimble competitors responding to rapidly changing markets, shifts in traditional supply chains and investor pressure for better earnings, productivity improvements and market leadership are a core component of this strategic assessment. It stresses the translation of strategic business objectives and models into an information systems architecture that combines data process, workflow, financial and simulation models. Managers need to understand how advanced technology creates more robust, scalable and adaptable information systems for the organization dedicated to continuous improvement and innovation. Lastly, the importance of making the business tolerant and immune to shifts in technology and able to leverage modern and up-to-date strategies and approaches in e-business environments cannot be overemphasized.

Recommended Readings

Bosworth, S. & Kabay, M. E. (2002). *Computer Security Handbook, (4th ed.).* New York: John Wiley & Sons.

Power, R. (2002). *Computer Security: Issues and Trends.* 2002 CSI/ FBI Computer Crime and Security Survey.

Stallings, W. (2000). *Network Security Essentials: Applications and Standards.* Upper Saddle River, NJ: Prentice-Hall.

Wack, J., Cutler, K. & Pole, J. (2002). Guidelines on Firewalls and Firewall Policy, Computer Security Division, Information Technology Laboratory, National Institute of Standards and Technology. Gaithersburg, MD: National Institute of Standards and Technology. Found online at: http://csrc.nist.gov/publications/nistpubs/800-41/ sp800-41.pdf.

Chapter IX

Security and the Importance of Trust in the Australian Automotive Industry

Pauline Ratnasingam
Department of Computer Information Systems
Central Missouri State University, USA

Biography

Dr. Pauline Ratnasingam received her Bachelor's in Computing from Monash University, Australia, and her PhD from Erasmus University, The Netherlands. In 1998, she was invited by KPMG Norlan & Norlan Institute, Australia, to participate in a nationwide survey examining the extent of e-commerce adoption in Australia and New Zealand. She played the role of consultant and researcher by contributing to sections of the survey questionnaire relating to e-commerce security and trust, analyzing the survey's findings, and designing the final report along with a group of consultants at KPMG.

She teaches courses on project management, management of IS, and e-commerce. She is an associate member of the Association of IS, and is a member of the Information Resources Management Association and Academy of Management. She has published articles and refereed journals on Internet-based B2B E-Commerce and trust. She has received a grant from the National Science Foundation to pursue her work on inter-organizational trust in B2B e-commerce.

Executive Summary

E-commerce–which is the sharing of business information, maintaining business relationships, and conducting business transactions by means of telecommunication networks–is growing at an exponential rate. The hype and growth of the Internet has attracted the growth of online B2B relationships. Internet business, in the United States alone, is forecast to increase as high as $7.3 trillion in 2004 (Gartner Group, 2000). Alternatively, the spatial and temporal separation between business partners generates an implicit uncertainty around online transactions. Uncertainties may arise when trading partners encounter barriers in communication (such as incompatible e-commerce systems, or lack of uniform standards) that may lead to conflicts. E-commerce adoption, unlike traditional information systems adoption, demands high levels of negotiation, cooperation and commitment from participating organizations. Selecting transaction sets, negotiating legal matters and defining performance expectations can burn up hours of staff time and also demand financial and technological resources. Managers have often cited a lack of trust as the main reason for failed business relationships. O'Hara-Deveraux and Johansen state that "trust is the glue of global work space and technology does not do much to create relationships" (1994, pp. 243-244). One way to ensure security and success of e-commerce is to establish trustworthy business relationships. This chapter aims to discuss the importance of trustworthy business relationships as a means to mitigate risks in EDI in the Australian automotive industry.

Introduction

Inter-organizational systems such as Electronic Data Interchange (EDI) have been the main form of business-to-business e-commerce application in the automotive industry for the last three decades. Previous studies in EDI adoption examined its competitive advantages from an environmental, organizational and technological perspective. This study draws upon insights developed within the sociology of technology, in which innovation is not simply a technical-rational process of "solving problem," but also involves technical, political and behavioral perspectives required for building inter-firm trust. The transition to cooperative relationships between buyers and suppliers may be more difficult for automotive companies due to high levels of complexity, compatibility, long lead times and ingrained adversarial supplier relationships of the past (Langfield-Smith & Greenwood, 1998). Further, security has become an important issue because EDI and most e-commerce systems do not operate unilaterally. For example, the Japanese automotive companies have a long-established history of developing relationships with their suppliers based on dependence and cooperation. Moreover, unlike the Japanese, in western countries like Australia, cooperative partnerships are a relatively recent phenomenon, and may be a distinct contrast to the adhoc relationships of the past.

We describe the Australian automotive industry followed by a discussion of the security issues from three perspectives: namely technical, political and behavioral. We then discuss the importance of trust among business partners in the automotive industry and conclude the chapter with implications for businesses.

Background Information of the Australian Automotive Industry

The Australian automotive industry has a well-developed supplier strategy which began from traditional EDI applications via value-added-networks to Internet-based EDI and today e-marketplaces (e.g., Covisint.com). E-marketplaces involve many buyers and suppliers trading over the Internet. The automotive industry was the first Australian industry to introduce EDI on a coordinated industry-wide basis and has been using EDI since electronic data transmissions commenced in the mid-1980s. EDI was used as a medium to communicate and transact production requirements for the five car manufacturers (namely Ford, General Motors Holden, Toyota, Mitsubishi and Nissan) to their component suppliers.

Security Issues

Security is still a number one concern for businesses that want to adopt EDI or e-commerce. Many businesses perceive that e-commerce transactions are insecure and unreliable. Despite the assurances of technological security mechanisms (such as encryption and authorization mechanisms, digital signatures and certification authorities), trading partners in business-to-business e-commerce do not seem to trust the personnel involved in the transactions. In this section we discuss security risks from three perspectives–namely technical, political and behavioral perspectives.

Technical Perspective

Business partners are subject to security attacks and intrusions by hackers. The security break-ins not only result in revenue losses for businesses, but also result in projecting adverse perceptions of e-commerce security. The information transmitted may be vulnerable at various points, including the trading partner's in-house applications, interface, translation software, network connection or communication management, as well as the carrier's network and mailbox services. The widespread use of Internet-based EDI has not only changed the way business is conducted, but has also introduced new risks that need to be addressed. The Internet, originally designed for scientific research use, emphasizes open communication and has many inherent security flaws. For example, Internet-based EDI security is still an administrative nightmare with problems from eavesdropping, password sniffing, data modification, spoofing and repudiation (Bhimani, 1996; Drummond, 1994). Other e-commerce risks include snooping, misuse, theft, corruption of information, theft of identity and personal threats. Cross-vulnerabilities that exist between interdependent trading partners in an e-commerce network can put organizations at risk due to the "domino effect" of one trading partner's security failure comprising the integrity of the other trading partner's system (Jamieson, 1996; Marcella et al., 1998). Given the computer-dependent nature of the automotive industry, it is important for business partners to build and maintain trustworthy relationships in order to mitigate risks from a technical perspective.

Political Perspective

Power, which is "the capability of a firm to exert influence on another firm to act in a prescribed manner," is an important contextual factor in EDI adoption, because of its influence.

Previous studies in the automotive industry suggest that Ford applied power when their EDI network was introduced (Webster, 1995). Ford's main objective was to gain competitive advantage by locking their suppliers into their system, and their competitors out of them. Ford made it clear to their established suppliers that they should use EDI. Although Ford did provide their suppliers with initial training and software to run on IBM machines, suppliers with incompatible systems or with no systems were requested to find appropriate solutions as quickly as possible. Clearly, this was a situation where coercive power exercised by Ford was seen in establishing connections that involved the expense of the suppliers buying new equipment. Examples of coercive sources of power an automotive manufacturer may exercise include, slow delivery on vehicles, slow payment on warranty work, unfair distribution of vehicles, turndowns on warranty work, threat of termination and bureaucratic red tape. It is here where trust can develop. Ford can either choose to see it proactively and renew their suppliers' contract or choose to punish their suppliers by terminating their contracts. The absence of collaboration or prior consensus about the structure, function and design of these networks provided suppliers with few opportunities to develop their knowledge and expertise in EDI use (Ratnasingam, 2000).

Similarly, Hart and Saunders (1997) suggest that EDI adoption was due to the pressure from the more powerful trading partners, usually buyers. Their findings indicated that power was negatively related to the volume of EDI transactions, reflecting that while electronic networks may facilitate easier exchanges, they may not necessarily lead to increases in the frequency of business transactions. EDI not only affects the efficiency of coordination, but also the power dependency and structural aspects of inter-organizational relationships. Thus, power exists on two levels: (1) as a motive, and (2) as a behavior.

Behavioral Perspective

The automotive industry is centered on assembling motor vehicles that involved standardized routine business operations. Buyers observe their supplier's behavior in their daily transactions, communications and business operations. Over a period of time, buyers are able to predict the performance of their suppliers. Ford possessed a set of strict measures that they applied when their suppliers did not cooperate. Their suppliers' competencies, product quality, timeliness of delivery, service quality and how they resolved disputes were observed. The supplier performance checklist determined whether to renew the contracts of their suppliers.

"*Our suppliers do have to meet the standards outlined in the Suppliers' Performance Assessment. Although, our suppliers have been trading with us for a long time, we usually undertake a screening test to examine their credibility, technical ability and skills. A standard of 85% and above was expected in their performance." — Ford EDI Coordinator*

Lessons Learned and the Importance of Trust

Previous research suggests that one key barrier to successful e-commerce adoption is the lack of trading partner trust, mainly derived from uncertainties, lack of open communications and information sharing (Cummings & Bromiley, 1996; Doney & Cannon, 1997; Ganesan, 1994; Gulati, 1995). Despite the assurances of technological security mechanisms, trading partners do not seem to trust the "people side" of the transactions.

Mayer, Davis and Schoorman (1995:712) defined trust as:

"...the willingness of a party to be vulnerable to the actions of another party based on the expectation that the other will perform a particular action important to the trustor, irrespective of their ability to monitor or control that other party."

The findings of this study indicated that trust is important in the automotive industry as business partners need to cooperate, collaborate and communicate timely and relevant information, in order to facilitate EDI that entails not only technological proficiencies, but also trust between trading parties, so that business transactions are sent and received in an orderly fashion. In the Japanese automotive system, the suppliers share an integral part of the development process, and they are involved early, assume significant responsibility, communicate extensively and directly with the production and process engineers (Dyer & Ouchi, 1993). Japanese suppliers are more cooperative and are willing to take risks. Hence, trust leads to cooperative relationships derived from commitment, frequent planned communications, and reduced transaction costs that eliminate inter-firm inefficiencies.

The findings of an exploratory study in three EDI organizations in the automotive industry identified two types of trust. The first type of trust (Soft Trust) is trust in their trading partner relationship (that is between a manufacturer and a supplier), experienced as follows:

Increased Communication During Initial EDI Adoption

"Although EDI was established in mid 1980s, to reflect back on our initial implementing procedures, we would still print off the order, and fax the same order again. Furthermore, after sending the order via EDI, we would call our suppliers to check if they had received it.

Hence, in the early stages of EDI adoption and implementation, we relied heavily on the daily audit trail, and other feedback mechanisms such as fax and telephone." — Ford Accounting Manager

Sharing and Understanding

"We met once bi-monthly to discuss business issues relating to EDI use within the automotive industry. We operate as a family unit (Closed User Group) and cooperate for the smooth flow of EDI operations, as we represented the automotive industry." — Ford Project Leader.

"Trust is related to security. Our trading partners respect the privacy and confidentiality of EDI messages, clearly outlined in the trading partner agreement. Our customers require the components on time with the right quantity, quality, and [it] has to be cost effective. Hence, we trust our trading partners (customers) to use procedures that to reduce uncertainties." — PBR, EDI Coordinator

"We do not only communicate via EDI, but other means such as telephone, fax and e-mail when there is a discrepancy. This related to communication openness, information sharing and concern. We do not check the delivery of goods, due to consistency in the quality service provided by our suppliers. Prior history of trading partner relationships enables us to make predictions about our suppliers' performance." — Ford Supply Chain Management Materials Planning and Logistics Core Group Manager

Belief in Your Trading Partners That They Will Perform the Required Task

The history of the trading partner relationship has shown that EDI trading partners do maintain a stable relationship.

"We believe that our trading partners' are competent enough to perform the task as required by them. This is based on our long-term trading partner relationship, as we have been trading with them for more than twenty years. Past experience has provided us knowledge that enabled us to predict their present and future business relationships. Initially we used to obtain wrong messages. Our trading partners have since shown competence in correctly and effectively performing the tasks." — Ford IT Manager.

The second type of trust (Hard Trust) focuses on the transmission medium (the technology). It is more specific and relates to integrity issues in the IT infrastructure in EDI/VANs and Internet-based EDI applications. Technology compatibility and organizational readiness to adopt EDI was seen to be important. The Federal Chambers of Automotive Industry committee manages the industry with members including nominees of the four Australian care manufacturers–Ford, Holden, Mitsubishi and Toyota. The Federation of Automotive Product Manufacturers (FAPM), importers and suppliers were also involved in this project. The mission of the committee is to establish and govern a reliable and secure communication network and building trustworthy trading partner relationships capable of hosting applications of e-commerce and business-to-business transactions for the Australian automotive industry. Hence, beyond the apparent need to develop cooperative relationships, trading partners formed governance and structural mechanisms that brought about repeated encounters, and used the passage of time to their advantage to build trust. Although, these trust-developing mechanisms come from EDI, they had a lot to do with trading partner interactions in the form of open communications and information sharing.

Conclusion

Australia's automotive industry is moving closer to developing one of the largest global forms of trading (i.e., via covisint.com). General Motors, Ford, Daimler Chrysler, Renault and Toyota have set up Covisint – a single giant trade exchange platform. Others who have joined Covisint included; Mazda, Volvo, Jaguar, Land Rover (via Ford), Fiat, Suzuki, Isuzu and Subaru (via General Motors), Nissan (via Renault), and Mitsubishi, Hyundai and Kai (via Daimler Chrysler). The Internet portal allows suppliers to bid for contracts, and the system streamlines the whole supply chain process, including procurement, planning, scheduling, manufacturing, and logistics for both sides of the transactions. In order for business partners to sustain long-term relationships, trust becomes important. Firms that trust their business partners are capable of making effective decisions. According to a recent article in the *Harvard Business Review* (2003) titled "Do not trust your intuition," it was found that although 45% of corporate executives rely on their intuitions to make business decisions, they agreed that the complexities and proliferation of e-commerce technologies have in turn made effective decision making the number one problem for management today.

Recommended Readings

Bhimani, A. (1996). Securing the commercial Internet. *Communications of the ACM*, *39*(6), 29-35.

Cummings, L.L. & Bromiley, P. (1996). The organizational trust inventory (OTI): Development and validation. In Kramer, R.M. & Tyler, T.R. (Eds.), *Trust in Organizations: Frontiers of Theory and Research* (pp. 302-320). Thousand Oaks, CA: Sage Publications.

Doney, P.M. & Cannon, J.P. (1997). An examination of the nature of trust in buyer-seller relationships. *Journal of Marketing*, (April), 35-51.

Drummond, R. (1995). Safe and secure electronic commerce. *Network Computing*, *7*(19), 116-121.

Dyer, J.H. & Ouchi, W.G. (1993). Japanese style partnerships: Giving companies a competitive edge. *Sloan Management Review*, (Fall), 51-63.

Ganesan, S. (1994). Determinants of long-term orientation in buyer-seller relationships. *Journal of Marketing*, 58(April), 1-19.

Gulati, R. (1995). Does familiarity breed trust? The implications of repeated ties for contractual choice in alliances. *Academy of Management Journal*, *38*(1), 85-112.

Hart, P. & Saunders, C. (1997). Power and trust: Critical factors in the adoption and use of electronic data interchange. *Organization Science*, *8*(1), 23-42.

Helper, S. (1991). How much has really changed between U.S. automakers and their suppliers? *Sloan Management Review*, *32*(4), 15-28.

Jamieson, R. (1996). Auditing and electronic commerce. *EDI Forum*, Perth, Western Australia.

Langfield-Smith, K. & Greenwood, M. R. (1998). Developing co-operative buyer-supplier relationships: A case study of Toyota. *Journal of Management Studies*, *35*(3), 331-353.

Marcella, A.J., Stone, L., & Sampias, W.J. (1998). *Electronic Commerce: Control Issues for Securing Virtual Enterprises*. The Institute of Internal Auditors.

Mayer, R.C., Davis, J.H. & Schoorman, F.D. (1995). An integrative model of organizational trust. *Academy of Management Review*, *20*(3), 709-734.

O'Hara-Devereaux, M. & Johansen, B. (1984). *Global Work: Bridging Distance, Culture, and Time.* San Francisco, CA: Jossey-Bass Publishers.

Ratnasingam, P. (2000). Power among trading partner trust in electronic commerce. *Internet Research: Electronic Networking Applications and Policy, 10*(1).

Webster, J. (1995). Networks of collaboration or conflict? Electronic data interchange and power in the supply chain, *Journal of Strategic Information Systems*, *4*(1), 31-42.

Chapter X

E-Commerce Security Planning

Daniel L. Ruggles
Vice President, Technology
Consulting Associates, LLC
USA

Biography

Daniel L. Ruggles is Vice President, Technology Consulting Associates, LLC. He is a Certified Information Systems Security Professional (CISSP). TCA, headquartered in Atlanta, Georgia, provides expert technology and business consulting services in a wide range of industries. For further information, call 404-303-1795 or visit the company's Web site at http://www.tca-llc.com.

Executive Summary

Multiple levels of potential security risk affect all the elements of an e-commerce site. Truly complete security protects three areas: (1) internal network and application services, (2) perimeter network access and application services, and (3) external network and services. To minimize security risks in these areas as well as to raise overall confidence in the e-commerce service, businesses must address problems with practical solutions involving privacy and security. Establishing an *e-commerce trust infrastructure* encompasses confidentiality, data integrity, non-repudiation and authentication. Striking the right balance between functionality and accessibility in e-commerce demands technical security measures. It also requires management vigilance with non-technical measures such as routine assessments of vulnerabilities, policies, education and a focus on making processes and policies easy to understand and simple to implement and monitor compliance.

Introduction

The design and architecture of an e-commerce site can take on many different appearances, depending on the company's objective and the type of commerce being conducted at the site. Unfortunately, many architectural decisions are not necessarily based on security considerations, causing potential holes in the overall network environment. E-commerce sites contain multiple levels of potential security risk affecting all their common building blocks: Web servers, commerce servers, database servers, routers, firewalls, network connectivity and the use of DMZs (demilitarized zones). Identifying and protecting sensitive data at all these levels can be accomplished by using various security features and building layers of complexity, commensurate with the value of the commerce being conducted.

Prevent the Past from Becoming the Future

Discussion of Issues

Based on responses from 503 computer security practitioners in U.S. corporations, government agencies, financial institutions, medical institutions and universities, the findings of the 2002 *Computer Crime and Security Survey* confirm that the threat from computer crime and other information security breaches continues unabated and that the financial toll is mounting. Highlights of the survey include:

- 90% of respondents (primarily large corporations and government agencies) detected computer security breaches within the last 12 months.
- 80% acknowledged financial losses due to computer breaches.
- 44% were willing and/or able to quantify their financial losses, which amounted to $455,848,000.
- As in previous years, the most serious financial losses occurred through theft of proprietary information (26 respondents reported a total of $170,827,000) and financial fraud (25 respondents reported a total of $115,753,000).
- For the fifth year in a row, more respondents (74%) cited their Internet connection as a frequent point of attack than cited their internal systems as a frequent point of attack (33%).
- 34% reported the intrusions to law enforcement. (In 1996, only 16% acknowledged reporting intrusions to law enforcement.)

Once again, to prevent the past from becoming the future, truly complete security must protect three areas:

1. *Internal network and application services.* Typically this is the internal network infrastructure, designed to support transfer and manipulation of internal business information. It consists of internal business application servers, workstations, printers and internal network access devices.
2. *Perimeter network access and application services.* This is the bridge between internal and external systems and typically acts as the "traffic cop" allowing or disallowing entry into the internal network.

Perimeter defense security is usually in the form of firewalls, proxy servers, intrusion detection systems, VPN and other similar services. When properly configured, perimeter defense security models prevent or detect attacks and reduce the risk to critical back-end systems from external attacks. Most companies and the trade press focus on this type of security, largely ignoring threats that come from within the company.

3. *External network and services.* Generally referred to as the Internet, this network provides access into suppliers' networks or customer access to the company's network. It is an ungoverned and unprotected network with many access points.

External network security services take on an added dimension with e-commerce. A truly staggering number of small e-commerce ventures don't employ secure Web pages (SSL) at checkout. And then there are the sites that obtain orders using a fully secured, 128-bit encrypted connection, but then e-mail the orders to the storeowner without any encryption! This is inherently dangerous, since e-mail messages are easy to monitor and are generally sent in clear text format. It is also dishonest to the customers who believe their secure transaction is being maintained.

Building an E-Commerce Trust Infrastructure

Implications

To take advantage of the opportunities of e-commerce, while minimizing the risks of communicating and transacting business online, every business must address practical problems and questions involving privacy and security, ever striving to increase the overall confidence in the underlying features of the system. Establishing an *e-commerce trust infrastructure* encompasses:

Authentication: Customers must believe and be assured that they are doing business with and sending private information to a real entity—not a "spoof" site masquerading as a legitimate bank or e-store.

Confidentiality: Sensitive Internet communications and transactions, such as the transmission of credit card information, must be kept private.

Data integrity: Communications must be protected from undetectable alteration by third parties during transmission on the Internet.

Non-repudiation: It should not be possible for a sender to reasonably claim that he or she did not send a secured communication or did not make an online purchase.

The solution for meeting these goals includes two essential components: (1) digital certificates for Web servers–to provide authentication, privacy and data integrity through encryption–and (2) a secure online payment management system–to allow e-commerce Web sites to securely and automatically accept, process and manage payments online. Together, these technologies form the essential trust infrastructure for any business that wants to take full advantage of the Internet.

The Proper Balance

Solutions and Recommendations

In addition, striking the right balance between functionality and accessibility is a critical facet of IT security supporting e-commerce. It involves six basic steps:

1. **Conduct routine assessments of vulnerabilities.** Routine means going beyond perimeter technology assessments and putting a heavy emphasis on the internal network and application services. Vulnerability scanners that perform the basic scans to examine vulnerabilities in computing platforms can be obtained at little or no cost. Fix the vulnerabilities within the enterprise! Every major software vendor has extensive information on how to secure an operating system, depending on the use. An example of this is securing Windows NT and IIS: http://www.microsoft.com/technet/treeview/default.asp?url=/technet/security/tools/chklist/wsrvsec.asp. (Refer to the list of recommended readings for other resources.)
2. **Write a policy that is clear, concise, relevant, up-to-date and maintainable.** Adhere to policies and educate users while maintaining them. Without a policy, there is no set of standards upon which to measure company activities. It is equally important that a policy contain more than platitudes. If it has no teeth, then why bother writing a policy that will never be enforced? People will do what is *inspected* and not necessarily what is *expected* from vague, generalized policy statements.
3. **Develop a set of minimum security best practices for implementation on all platforms.** This includes desktops, servers, routers, firewalls, applications (e-mail, Web servers, etc.). Relying solely on a secure perimeter without hardened systems leads to a false sense of security. Implementing an e-mail server, without making it a hardened system for an e-mail service, is just asking for trouble. It takes little time and effort to harden an e-mail server. If everyone has a Windows workstation, when was the last time any security patch was installed and how many varieties of Windows does the company use?

4. **Educate employees.** Teach developers how to incorporate sound security practices into applications. Teach end-users the "do's and don'ts" of good security. Post-It Notes with the password on the side of a terminal or under a keyboard, for example, are unacceptable. Strive to keep staff informed! Education and awareness are two of the least expensive ways to mitigate enterprise risk. Everyone likes to poke fun at Microsoft as it ventures into making its products more security aware. But how many companies write custom code or buy application packages without a single thought as to how secure that code really is?
5. **Review company processes.** Make sure that processes exist where they are needed. Does a process exist for system backup? What about protecting those backups? What about moving those backups off-site? If those processes exist, are they regularly tested? Implementing a SAN (storage area network) or any other new storage technology does not eliminate the need for some simple testing. It is too late to find out that all those backups or mirrored files are worthless when the need to restore arises. Test...test...test!
6. **Streamline processes.** Processes that are unpleasant to perform or feel unnecessary to the employees are less likely to be followed—despite their importance in ensuring security. Although security processes may never be fun, they should be easy to follow.

The purpose of good security architecture is to keep a network and its computers secure. It is not supposed to obstruct the course of doing business, nor is it intended to reduce the functionality of the tools needed to perform a job. A good security system does not work against company processes. It should mesh with current computer usage and system design, benefiting the enterprise as a whole.

Patrice Rapalus, Director of the Computer Security Institute (CSI), remarked with respect to the findings of the 2002 Computer Crime and Security Survey,

"Over the seven-year life span of the survey, a sense of the 'facts on the ground' has emerged. There is much more illegal and unauthorized activity going on in cyberspace than corporations admit to their clients, stockholders and business partners or report to law enforcement. Incidents are widespread, costly and commonplace. Post-9/11 there seems to be a greater appreciation for the significance of information security, not only to each individual enterprise but also to the economy itself and to society as a whole. Hopefully, this greater appreciation will translate into increased staffing levels, more investment in training and enhanced organizational clout for those responsible for information security."

An Inclusive Model

Over the last few years, more and more businesses have moved to the e-commerce delivery model, not to replace their "bricks-and-mortar," but to supplement it. In this evolution, the early focus on a high "cool factor" has been replaced by sound, practical marketing and sales strategy incorporating light graphics, speed, knowledge resources, visibility, quick checkout and customer service. The only development that has fallen behind is security. Sound security posture in a constantly changing "threatscape" requires continual vigilance and people who are motivated, alert, aligned with the mission and technologically current. Security is only as good as the weakest link. An adequate security model, therefore, must include elements that go beyond perimeter security to consider internal and external factors, as well as people and processes.

Recommended Reading

The Center for Internet Security. Available online: www.cisecurity.org/. (The Center's mission is to help organizations around the world effectively manage the risks related to information security. CIS provides methods and tools to improve, measure, monitor and compare the security status of your Internet-connected systems and appliances, plus those of your business partners. CIS is not tied to any proprietary product or service. It manages a consensus process whereby members identify security threats of greatest concern, then participate in development of practical methods to reduce the threats. This consensus process is already in use and has proved viable in creating Internet security benchmarks available for widespread adoption.)

Computer Crime and Security Survey. Available online: http://www.gocsi.com/press/20020407.html.

In Pursuit of the User Friendly, Impenetrable, Tamperproof, Impregnable Firewall Gelb Organization. (1998). Available online: http://www.gelb.com/chapter.htm.

Intruder Detection Checklist. Available online: http://www.cert.org/tech_tips/intruder_detection_checklist.html.

Network vs. Host-Based Intrusion Detection: A guide to Intrusion Detection Technology. Available onlin: http://secinf.net/info/ids/nvh_ids/.

Pace, N. (1998, November). *What Firewalls Can (and Can't) Do For You.* Available online: http://www.ne-dev.com/ned-02-1998/ned-02-firewall.html.

Ranum, M. J. *Intrusion Detection: Challenges and Myths.* Available online: http://secinf.net/info/ids/ids_mythe.html.

San Francisco State University - College of Extended Learning. Available online: http://msp.sfsu.edu/instructors/jlevin/security/security_toc.html.

Schupp, S. *Limitations of Network Intrusion Detection.* Available online: http://www.sans.org/infosecFAQ/intrusion/net_id.htm.

Shipley, G. *Watching the Watchers: Intrusion Detection.* Available online: http://www.networkcomputing.com/1122/1122f3.html.

State of the Practice of Intrusion Detection Technologies. Available online: http://www.sei.cmu.edu/publications/documents/99.reports/99tr028/99tr028exsum.html.

Stateful Inspection Firewall Technology: Tech Note. (1998). Available online: http://www.checkpoint.com/products/technology/stateful1.html.

UNIX Configuration Guidelines. Available online: http://www.cert.org/tech_tips/unix_configuration_guidelines.html.

UNIX Security Checklist v2.0. Available online: http://www.cert.org/tech_tips/unix_security/checklist2.0.html.

Wack, J. & Carnahan, L. (1995, February). *Keeping Your Site Comfortably Secure: An Introduction to Internet Firewalls.* NIST Special Publication 800-10. Avaiable online: http://csrc.ncsl.nist.gov/nistpubs/800-10/main.html.

Windows NT Configuration Guidelines. Available online: http://www.cert.org/tech_tips/win_configuration_guidelines.html.

Section IV

Glossary and Index

Glossary of Terms

Biometric Controls

Biometric controls are types of authentication devices used to confirm an individual's identity. These controls are based on unique biological, behavioral and physical characterisitcs such as voice or fingerprint.

Cookie

A website (server) places one or more cookies on a hard disk to identify a user for future reference. A cookie requests that the user store the information (i.e., a website can remember information about a user). A cookie can monitor your actions on particular websites and record your preferences. The information stored by a cookie can be kept on the user's computer. It is possible to view the cookie on your hard disk.

Digital Certificate

A digital certificate is a digital file issued to an individual or company by a certifying authority that contains the individual's or company's public encryption key and verifies the individual's or company's identity. (Turban et al., 2000, p. 507)

Digital Watermark

A pattern of bits inserted into a digital image, audio or video file that identifies the file's copyright information (author, rights, etc.). The purpose of digital watermarks is to provide copyright protection for intellectual property that is in digital format. (webopedia.com)

Disk Operating System

Disk Operating System (DOS) refers to any operating system, but it is most often used as shorthand for MS-DOS (Microsoft Disk Operating Systems). Originally developed by Microsoft for IBM, MS-DOS was the standard operating system for IBM-compatible personal computers. (www.webopedia.com, 2003)

Electronic Commerce

Electronic commerce (e-commerce) is buying and/or selling products online. The payments are often in the form of electronic means. E-commerce can be business to business (B2B) or business to consumer (B2C).

Encryption

Encryption is a process of making messages indecipherable except by those who have an authorized decryption key. (Turban et al., 2000, p. 507)

Firewall

A firewall prevents unauthorized access to or from a private network, protecting the information and data in that network. Firewalls are installed in both the hardware and software to isolate a private network from intrusion from unauthorized public networks. All messages from an intranet pass through the firewall where any messages are reviewed. Those that fail to meet the security conditions for that network will be denied access.

Interoperability

Interoperability is the ability of a system to work with other systems to share information. The information shared allows both systems to operate more efficiently together.

Mobile Commerce

Mobile commerce (m-commerce) is any electronic transaction or information interaction conducted using a mobile device and mobile networks (wireless or switched public network) that leads to the transfer of real or perceived value in exchange for information, services or goods. (MobileInfo.com)

Public Key Infrastructure (PKI)

A public key infrastructure (PKI), also known as a trust hierarchy, allows Internet users to safely and securely exchange information and/or money via a trusted authority. PKI uses registration authorities such as digital certificates or certificate authorities to verify the individuals involved in an Internet transaction. Public key cryptography is the most common Internet method for verifying the message of a sender or encrypting a message. Two main types of cryptography are secret key and public/private key cryptography.

Secure Socket Layer

Secure Socket Layer (SSL) is an encryption method for transferring data between a client and a server. (Strategic Web Ventures.com)

Sniffer/Sniffing

A network manager uses a sniffer program to monitor network traffic. The obtained data is then analyzed for inefficiencies and/or problems on the network that can later be fixed by the network manager. A sniffer can also detect and read any information traveling along a network. A sniffer can be used to steal information from a network, putting a network at a security risk. For example, a sniffer program can find and share passwords.

Spam (or Spamming)

Spam is unsolicited mail, sent to a large number of e-mail addresses, often in the form of an ad or a scam. The sender views spam as a type of bulk mail obtained, for example, from a mailing list. Serious problems due to spam include: wasting bandwidth, wasting storage space and wasting time.

Spoofing

Spoofing is the creation of TCP/IP packets using somebody else's IP address. Routers use the "destination IP" address in order to forward packets through the Internet, but ignore the "sourceIP" address. That address is only used by the destination machine when it responds back to the source. (Internet Security Systems)

References

Internet Security Systems (1994-2003). Retrieved July 29, 2003, at: http://www.iss.net/security_center/advice/Underground/Hacking/Methods/Technical/Spoofing/default.htm.

MobileInfo.com (2001-2003). Accessed July 29, 2003, from http://www.mobileinfo.com/Mcommerce/.

Strategic Web Ventures (2003). Accessed July 28, 2003, at: http://www.strategicwebventures.com/definitions/Glossary/SSL/.

www.webopedia.com (2003). Retrieved July 28, 2003.

Index

S

T

U

V

W